PROGRAMMING WI....

PROGRAMMING WITH OPENSCAD

A Beginner's Guide to Coding 3D-Printable Objects

by Justin Gohde and Marius Kintel

no starch press

San Francisco

Printed in the United States of America

First printing

25 24 23 22 21   1 2 3 4 5 6 7 8 9

ISBN-13: 978-1-59327-954-7 (print)
ISBN-13: 978-1-59327-955-4 (ebook)

Publisher: William Pollock
Production Manager: Rachel Monaghan
Production Editor: Dapinder Dosanjh
Developmental Editor: Jill Franklin
Cover Illustrator: Gina Redman
Interior Design: Octopod Studios
Technical Reviewer: Miro Hrončok
Copyeditor: Sharon Wilkey
Compositor: Maureen Forys, Happenstance Type-O-Rama
Proofreader: Emelie Battaglia
Indexer: Beth Nauman-Montana

The following images are reproduced with permission: Figure 1 in the introduction was created by Gustavb and is licensed under the Creative Commons Attribution-Share Alike 3.0 Unported license. The photo of the Leaning Tower of Pisa featured in Figures 7-1 to 7-4 is by Svetlana Tikhonova. The image is covered by the CC0 1.0 Universal (CC0 1.0) Public Domain Dedication license.

For information on book distributors or translations, please contact No Starch Press, Inc. directly:
No Starch Press, Inc.
245 8th Street, San Francisco, CA 94103
phone: 1.415.863.9900; info@nostarch.com
www.nostarch.com

Library of Congress Cataloging-in-Publication Data
LCCN: 2020943329

About the Authors

Justin Gohde has been teaching math and computer science for more than 20 years. He is the head of computer science at Trinity School in Manhattan, New York, where he oversees the computer science curriculum and programs, including the Design Lab makerspace, which features a wide collection of rapid prototyping tools for supporting computer science, robotics, and other STEAM-focused projects.

Marius Kintel is the primary author and maintainer of OpenSCAD. He is a software engineer with more than 20 years of professional experience in diverse fields such as 3D visualization, physical computing, manufacturing automation, and cloud computing. In 2007, he became involved in the RepRap project out of an interest in learning electronics and the opportunity to transfer his knowledge of 3D graphics into the creation of tangible objects. Together with the rest of the local RepRap team at the Metalab hackerspace in Vienna, Austria, he created OpenSCAD out of a need for an open source multiplatform design tool more suitable for 3D printing. The project was adopted by the maker community at large and has since grown to become one of the most popular open source 3D modeling tools for 3D printing.

About the Technical Reviewer

Miro Hrončok is a software engineer mostly working on integrating the Python ecosystem into Fedora Linux. Besides that, he also takes care of the 3D printing stack (including OpenSCAD) in Fedora. He teaches Python and 3D printing basics at the Faculty of Information Technology, Czech Technical University in Prague.

BRIEF CONTENTS

CONTENTS IN DETAIL

3
2D SHAPES

4
USING LOOPS AND VARIABLES

ACKNOWLEDGMENTS

First, the authors would like to acknowledge the hard work and significant efforts put in by so many folks operating behind the scenes to make this book happen. As technical reviewer, Miro Hrončok made many, many insightful and meaningful contributions to the evolution of each chapter. His feedback and advice were well seen, well considered, and well appreciated. We'd also like to thank everyone at No Starch for their remarkable patience, dedication, and guidance throughout the book writing process. In particular, we'd like to acknowledge and thank Bill Pollock for agreeing to publish one of the first books about OpenSCAD, Tyler Ortman for getting the ball rolling, and finally Jill Franklin and Dapinder Dosanjh for getting the project back on track and over the finish line.

Justin would like to thank the many Trinity students and faculty who helped prototype the core lessons and activities presented in this book—in particular, colleagues Noah Segal, Mark Schober, Rob Newton, Jean Kim, and Andrew Rose for their thoughtful contributions to the conversation. Much gratitude is owed to Chris Martin, for initiating scores of London chats on the pedagogical importance of using starting points to motivate self-reflective, inquiry-based learning. And, finally, an infinite debt of thanks is due to Yasuho Mori due to her eternal patience, persistent optimism, and steadfast support.

Marius would like to thank all contributors and users of OpenSCAD. In particular he would like to thank Torsten Paul, who has been instrumental in keeping OpenSCAD moving forward over a number of years.

Also, thanks to the Metalab hackerspace in Vienna, and all the diverse and colorful individuals that frequent the space, for creating an inspiring environment where projects like OpenSCAD can evolve. Special thanks go to Claire Wolf for single-handedly cranking out the initial codebase for OpenSCAD and Philipp Tiefenbacher for being an amazing project partner and visionary who helped bootstrap all the activities that eventually became OpenSCAD. And finally, to Jane Tingley for her continuous support, encouragement, and unwavering belief.

INTRODUCTION

Programming with OpenSCAD: A Beginner's Guide to Coding 3D-Printable Objects introduces the versatile, text-based OpenSCAD 3D CAD software. This book guides readers through using arithmetic, variables, loops, modules, and decisions to design a series of increasingly complex 3D designs, all of which are suitable for 3D printing.

What Is OpenSCAD?

OpenSCAD (pronounced *Open-S-CAD*) is text-based software for creating solid 3D models. It allows you to design these models by writing code, which gives you (the designer) full control over the modeling process and allows for easy changes to any part of your design throughout that process. OpenSCAD also makes it easy to create designs that are defined by configurable parameters, which means you can build designs with change in mind.

OpenSCAD is a *descriptive programming language*: its coding statements *describe* the shape, size, and combination of each component of your overall design. OpenSCAD provides two main 3D modeling techniques: you can create complex objects from combinations of simpler shapes (known as *constructive solid geometry*) or extrude 2D *.dxf* or *.svg* outlines into 3D shapes. Unlike most other free software for creating 3D models (such as Blender), OpenSCAD focuses on the mechanical, rather than the artistic, aspects of 3D design. Thus, OpenSCAD is the application to use when you're planning to create 3D-printable models, but probably not what you're looking for when you're more interested in making computer-animated movies.

OpenSCAD is free, open source software that's available to download for Windows, macOS, and Linux, as well as other systems. Specific system requirements are available at *https://openscad.org/downloads*. This book covers OpenSCAD 2021.01, which is the latest version at the time of writing.

Who This Book Is For

This book is for beginners who are either new to code, new to 3D design, or new to OpenSCAD. While no prior background with either coding or 3D design is necessary to access the material presented in this book, some experience with introductory algebra would be helpful. No particular math beyond basic arithmetic (adding, subtracting, multiplying, and dividing) is required. However, some previous experience using variables in simple equations would be a useful starting point, as would some experience graphing points on the xy-plane.

In line with our intentions to make this book beginner-friendly, we have deliberately chosen to focus on a subset of OpenSCAD. Our goal is to introduce its most useful elements in an accessible manner. In some cases, that means we briefly introduce a topic earlier in the book that we return to in more depth in later chapters. This spiraling is intended to help you form a strong foundation in the basics before adding layers of nuance. Our examples and projects have been curated to allow for maximum creative utility while still making the topic accessible to readers who are new to coding.

Why Learn to Code with OpenSCAD?

While learning to code can be fun and exciting, it can also be challenging for beginners to figure out the *where* and *why* behind the inevitable errors in their coding projects. In contrast to the opaqueness of other text-based programming languages (in which it's hard to see what's going on behind the scenes), OpenSCAD's visual nature gives users immediate feedback regarding the correctness of a particular approach.

Writing text-based code to create a 3D object is a powerful and effective way to learn how to structure long sequences of coding statements. Like more familiar imperative programming languages (JavaScript, Python, and so on), OpenSCAD includes variables and common control structures (such as if statements and loops) and supports the creation of programmer-defined modules and libraries. Additionally, OpenSCAD employs common syntax elements, such as using curly brackets ({ }) and semicolons (;), to organize statements as well as the familiar set of arithmetic operators and conventions. OpenSCAD not only opens up the world of text-based 3D design, but also teaches skills that are transferable to many other popular programming languages.

Learning to code with OpenSCAD also offers unique advantages for developing *computational thinking*. This computer-specific approach uses decomposition, abstraction, patterns, and algorithms to solve a problem in a way that makes it easy for a computer to carry out the solution. Developing an intuition for computational thinking can be difficult for beginners in other programming languages, but OpenSCAD makes it easy with algorithms and coding statements that literally take shape. Applying abstraction and patterns means visually identifying the repetitive and predictable elements in a design; decomposition becomes splitting a complex design into well-defined smaller pieces, and algorithms naturally extend from creating a list of steps that need to happen in order to create a design. The tactile feedback that comes from turning an OpenSCAD design into a physical 3D-printed object adds an entirely new dimension to learning to code.

STEM (science, technology, engineering, math) and *STEAM* (add art into the mix) are two recently popular acronyms that describe learning activities existing at the intersection of two or more of these traditionally separated disciplines. Learning to code with OpenSCAD is like taking a holistic, STEAM-based approach to learning how to code. OpenSCAD coding projects require translating visual shapes into concisely worded textual descriptions, and vice versa. Designs that start as hand-drawn sketches are converted to mathematical coordinate representation, with features that are estimated with proportionality. Designing with OpenSCAD code requires navigating both orthogonal and perspective views of 3D objects, and thinking about 3D shapes in terms of their 2D shadows. 3D-printing an OpenSCAD design develops engineering skills by requiring the consideration of physical tolerances and the adjustment of machine settings. In true STEAM fashion, this book asks you to simultaneously develop, combine, and practice skills typically relegated to the separate disciplines of technology, engineering, art, and math as you learn to code with OpenSCAD.

Learning to code with OpenSCAD has quite a few advantages:

- OpenSCAD is popular, free, and open source.

- OpenSCAD is easy to learn and uses a common and transferable, text-based syntax that is shared with other popular programming languages.

- Designing 3D objects with OpenSCAD preserves a discoverable design history. Unlike other 3D-design software, where clicking Undo removes a step, with OpenSCAD, you can easily modify earlier steps in the design process without erasing later ones.

- The compact size of text-based OpenSCAD files (*.scad*) makes sharing, storing, and modifying OpenSCAD models faster and more efficient than working with typical 3D-modeling file formats.

- OpenSCAD has an easy-to-find console window for immediate and easy debugging feedback.

- OpenSCAD coding projects are 3D-printable.

- OpenSCAD is an effective first programming language choice for visual learners.

- Learning to code with OpenSCAD builds a foundation in computational thinking while also reinforcing spatial and mathematical reasoning in an interdisciplinary, STEAM-based context.

3D Printing and OpenSCAD

Most people use OpenSCAD to create designs for 3D printing. At its core, *3D printing* is a tool for transforming virtual models into actual physical objects. OpenSCAD is a great choice of software to use when you're creating parts to manufacture with a 3D printer. However, access to a 3D printer is in no way a prerequisite for this book or for learning to use OpenSCAD. We certainly recognize the appeal of seeing and touching your 3D designs, so we've sprinkled 3D-printing tips throughout this book, anticipating that many readers will want to interact with their virtual designs in the real world.

3D printing is used in an ever-increasing number of areas: mechanical engineering, product design, animation, architecture, sculpture, medicine, geology, rocketry, and the list goes on. 3D printing first gained popularity for its uses in rapid prototyping, which allows designers to create physical models and receive real-world feedback much faster than previously possible. However, in addition to prototyping early versions of a design, 3D-printing technologies have advanced to the point where it's now possible to directly manufacture products in a variety of materials. Designers can now use 3D printing to build the final version of their design, using many types of plastic, glass, metal, magnets, cement, porcelain, bio-matter, and even edible foods! In fact, it's no longer unusual for mechanical engineers to 3D-print metal rocket-engine parts, for dentists to 3D-print porcelain dental implants, for architects to 3D-print residential houses in cement, or for sculptors and jewelers to 3D-print a wax base for a lost-wax casting.

Although many types of 3D-printing technologies exist, *fused-filament fabrication* remains the cheapest and most readily accessible technology available. Most of the 3D-printing tips in this book are best suited for fused-filament fabrication, which builds a 3D form by melting successive layers of plastic on top of each other.

What's in This Book

This book is split into three sections:

- Chapters 1 through 3 introduce how to draw and combine basic 3D and 2D shapes.

- Chapters 4 through 6 introduce loops, modules, and decisions so that you can add new layers of efficiency to your design process.

- Chapter 7 serves as a case study to reinforce prior topics and introduce higher-level design skills that work hand in hand with computational thinking.

A series of Design Time challenges accompany the first six chapters of the book. These exercises provide quick designs to replicate, suitable to the scope of each chapter's content. A small collection of Big Projects conclude each chapter. These projects, which require substantively more time and effort than the Design Time activities, are deliberately chosen to present a cumulative challenge.

The designs in both the Design Time and Big Project sections are presented without absolute coordinates, as they are intended to inspire you to build toward a general resemblance without focusing too much on details. For these exercises, the big details like proportionality and shape combinations matter more than anything else. All Design Time and Big Project exercises are well suited for 3D printing.

The following list gives a breakdown of the topics presented in each chapter:

Chapter 1: 3D Drawing with OpenSCAD
Introduces the OpenSCAD interfaces and teaches you to draw and place a few of the OpenSCAD primitive 3D shapes: cuboids, spheres, cylinders, and cones. OpenSCAD can also import 3D shapes generated by other applications, and we introduce that here as well. Another important concept covered is how to combine multiple shapes in a few ways. Finally, you'll learn how to export an OpenSCAD 3D design for 3D printing. The Big Projects in this chapter are designed to help you get to know the settings on your 3D printer's preparation software.

Chapter 2: More Ways to Transform Shapes
Presents a variety of additional transformation operations that can be applied to the 3D shapes introduced in Chapter 1. You'll learn how to rotate, mirror, and adjust the proportionality of 3D shapes. You'll also learn more sophisticated methods of combining shapes, including how to wrap a hull around two shapes and how to spread the properties of one shape along the edges of another shape with the minkowski operation. 3D-printing tips in this chapter introduce the concepts of infill and shell. The Big Projects ask you to combine multiple topics from Chapters 1 and 2 to produce objects you may actually use: a game die and a desktop organizer.

Chapter 3: 2D Shapes
Discusses an alternate way of approaching 3D design—building up a 3D form from its 2D *shadow*. You'll learn how to draw with primitive OpenSCAD 2D shapes, including circles, rectangles, polygons, and text (including emoji). You'll also learn how to combine those 2D shapes by

using most of the same operations you studied in Chapter 2, as well as a new 2D operation called *offset*. Finally, you'll see how to bring 2D shapes into the 3D world by extending them along the z-axis with a variety of new operations. 3D-printing tips in this chapter discuss resizing your 3D models for printing, including how to break a large model into multiple pieces so you can grow your 3D prints beyond the limited size of your 3D printer's build platform. The Big Projects include storytelling dice, a dice holder, and a 3D trophy built from a 2D profile.

Chapter 4: Using Loops and Variables

Introduces a new tool for computational thinking: the for loop. You'll learn how to use variables and for loops to repeat shapes. The best part is that you'll learn how to vary the characteristics of a shape (such as its size, position, or rotation) as it's repeatedly drawn by the loop. This chapter also introduces comments and console printing as useful tools for planning and debugging your designs. 3D-printing tips in this chapter relate to exploring a few gotchas that may surprise you when you try to create 3D-printed objects from OpenSCAD designs: the limitations of small-scale features, reconfiguring a design to avoid fusing together parts that are supposed to be separate, and breaking a design into separate *.stl* files to print different pieces with different-colored filament. The Big Projects include a detail test, a Towers of Hanoi game, and a tic-tac-toe game.

Chapter 5: Modules

Introduces yet another computational thinking tool: decomposing a design into multiple modules. You'll learn to use OpenSCAD modules to create your own shapes, as well as use separate files to group your new shapes into a reusable (and shareable) library. You'll also create and use parameters to control characteristics of your shapes, as well as define variables within modules so that updating the design of new shapes is quick and easy. The Big Projects in this chapter include a skyscraper module and a library of new LEGO brick designs.

Chapter 6: Dynamic Designs with if Statements

Introduces the if statement, which allows you to create dynamic designs that change according to a certain condition. You'll learn to create a variety of complex conditions using Boolean and logical operators, as well as extended if statements, and if...else statements. You'll automate some of the design configurations suggested in the Big Projects from Chapter 4, as well as incorporate random numbers to add fun and unpredictable elements to your design and make repeated elements more organic and natural. The Big Projects include creating a random forest, a clock face, and a city of random skyscrapers.

Chapter 7: Designing Big Projects

Presents a capstone project that walks through the process behind creating a big, multifile design. You'll apply formal characteristics of computational thinking by using the iterative design cycle to reinforce and expand the ideas presented in the first six chapters. You'll leverage the *walking skeleton* approach to evolve a simple version of the Leaning Tower of Pisa into a 3D model that bears a high resemblance to the actual tower. You can 3D-print this building as a trophy to congratulate yourself for all that you will have learned by following along with the material presented in the book.

If you get stuck on any exercise in this book, suggested solutions to the Design Time and Big Project exercises (along with all chapter examples) are available at *https://programmingwithopenscad.github.io/*.

Terminology and Conventions Used in This Book

Many introductory books on programming and computational thinking are available, and each author makes tough decisions as to how much granular detail is necessary for the audience they are trying to reach. As this book is meant for beginners, we have chosen to keep a high level of abstraction with regard to our vocabulary and conventions. Although some of the following terms have more precise definitions in other circumstances, our philosophy for this book is consistent with "don't sweat the small stuff."

We use the following vocabulary in the book:

Shape　Any graphical 2D or 3D object created by OpenSCAD.

Design　An OpenSCAD creation (that is, an OpenSCAD *program*), which usually consists of a combination of multiple shapes.

Operation　An OpenSCAD command that changes the appearance/ properties of one or more shapes.

Parameter　Any value that specifies characteristics of shapes, operations, modules, or functions.

Preview　The process of quickly displaying a design on-screen.

Render　The process of fully evaluating the geometry of a design (and showing it on-screen). Once it's rendered, you can export a design.

Units　All dimensions in OpenSCAD are specified in *units*. A unit is usually a millimeter (by 3D-printing industry convention), but OpenSCAD is technically unitless. All models should be explicitly sized in 3D-printing preparation software just prior to printing.

Width　The dimension associated with the x-axis, which is the "left-right" axis when 3D printing.

Length　The dimension associated with the y-axis, which is the "forward-backward" axis when 3D printing.

Height　The dimension associated with the z-axis, which is the "up-down" axis when 3D printing.

2D shapes　Shapes with a width and length, but no height.

3D shapes　Shapes with a width, length, and height.

A Brief Introduction to 3D Design with OpenSCAD

If you've never worked with virtual 3D models before, manipulating the 3D designs you create in this book via the use of a 2D computer screen can be confusing at first. Understanding some of the basics involved in creating the illusion of 3D space on a 2D surface can also help you navigate the transition to a 3D-modeling environment.

Understanding 3D Points

3D objects have a width, length, and height, so drawing a representation of 3D shapes requires the use of three separate axes: the x-axis, y-axis, and z-axis (Figure 1). The intersection of all three axes is called the *origin* and is indicated as the point (0, 0, 0) on the graph. Each axis proceeds in both positive and negative directions from the origin. Although a width, length, or height must be positive, the position of an object on a particular axis may be in the negative direction (which is relative to the location of the origin).

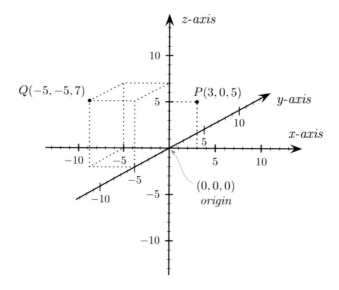

Figure 1: 3D coordinate system (Cartesian coordinate system. Created by Gustavb using PSTricks, licensed under the Creative Commons Attribution-Share Alike 3.0 Unported license: https://commons.wikimedia.org/wiki/File:Cartesian_coordinates_3D.svg)

Sometimes a 2D screen makes it hard to determine the exact 3D point being viewed. For instance, in Figure 1, the point (3, 0, 5) could also be interpreted as the point (0, 4, 3.5). When in doubt about the size or position of a particular shape, rotate your design to gain a fuller perspective of the feature. As you rotate your design, a miniature graph legend (circled in red in Figure 2) rotates accordingly to help you keep track of which axis is which.

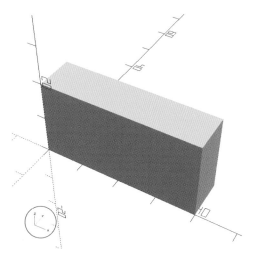

Figure 2: The graph legend keeps track of axis labels.

This legend is helpful because the axes in OpenSCAD aren't labeled. This graph legend is a good feedback tool for interpreting the changing orientation of the *width, length,* and *height* as you rotate your view to understand each part of your design.

Using the OpenSCAD 3D-View Toolbar

OpenSCAD uses a variety of perspectives and color shading (as does other 3D CAD software) to facilitate the representation of 3D shapes on a 2D computer screen. In addition to using a mouse, touchscreen, or trackpad to rotate your design, the OpenSCAD 3D-View toolbar (Figure 3) provides several buttons for quickly rotating the 3D view of your design to an orthogonal 2D view, which can help reveal a shape's true location and dimensions.

Figure 3: Quick 2D orthogonal views of a 3D shape

In order, the buttons reveal the following 2D views: right, top, bottom, left, front, and back.

Final Tips for Getting the Most Out of This Book

OpenSCAD has many more advanced features and capabilities than are included in this book. Consider these chapters a starting point in your exploration of the design possibilities offered by OpenSCAD. We've included an Afterword to provide a context for the development of OpenSCAD as an open source project, and to provide suggestions for further learning once you've finished reading the book. We encourage you to consult the documentation resources at *https://openscad.org/*, as well as the language reference (Appendix A) included in the back of this book to explore the full range of possibilities offered by the language. For a quick view of the basic features of OpenSCAD covered in the first four chapters of the book, we've also included a visual reference (Appendix B).

To truly learn how to design and code 3D-printable objects with OpenSCAD, you'll need to put the book down periodically. Give yourself an opportunity to type and modify our examples, as well as to create your own versions of our Design Time and Big Project exercises. Then, use the book as a starting point for designing and coding your own projects. In fact, once you've learned something new, take a break from the book. Remix or extend our projects and examples, or design something entirely new. Try to design something useful, something that will help you apply that new lesson to a project you're genuinely interested in. Show off and share your designs. Maybe even give your 3D-printed objects as a gift. Learning something new is much easier when you're genuinely engaged with the topic, so most of all, have fun!

1

3D DRAWING WITH OPENSCAD

This chapter introduces the OpenSCAD 3D design software with its own built-in programming language. You'll learn how to use text-based commands to draw the basic 3D shapes that will act as the building blocks for all the designs in this book. OpenSCAD's easy-to-learn programming language, specifically designed for 3D printing, is a descriptive language that offers a more natural way of describing geometry than traditional programs.

Why Use OpenSCAD?

OpenSCAD is an open source program that is freely available for download. It is one of the most widely used 3D design software applications in the maker community, and as a result, many online resources are available. OpenSCAD was built to enable nondesigners to easily create 3D models. It does not have a graphical user interface like Photoshop. Instead, you define your design with text-based code, which makes it easier to move around different parts, change earlier steps in the design process, share sections of your designs with other people, discuss your design problems in forums, and email designs to others. You can do similar things in OpenSCAD as are possible with other high-end tools; however, OpenSCAD is quick to learn, simple to use, and more accessible.

Getting Started with OpenSCAD

Creating a 3D design with OpenSCAD is a two-step process. First, in the Editor window, type a code statement to give OpenSCAD instructions about what to display. Figure 1-1 shows a code statement to draw a simple OpenSCAD shape circled in red.

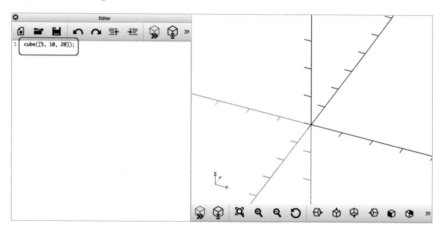

Figure 1-1: Code for a cube in the Editor window

This OpenSCAD code statement has two parts. The first part indicates the type of shape you want to draw (in this case, a cuboid). The second part, which contains what are called *parameters*, indicates the properties of that shape. Parameters allow you to specify values that modify the appearance of the shape. Parameters are always placed between parentheses ( ).

Next, draw your shape in the Preview window by clicking the **Preview** button (circled in red in Figure 1-2) to see a quick visual preview of your design.

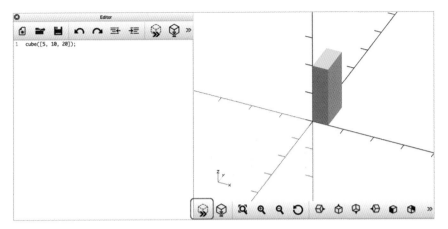

Figure 1-2: Drawing a cube after clicking the Preview button

Drawing Basic 3D Shapes

In this section, you'll learn how to write OpenSCAD code to draw cuboids (cubes or 3D rectangles), spheres, and cylinders, and you'll learn how to import shapes from other design programs.

Drawing Cuboids with cube

Use the cube command to create a cuboid (as shown in Figure 1-2):

```
cube([5, 10, 20]);
```

The first part of the statement, cube, indicates that you want to draw a cuboid. The parameters inside the parentheses modify the cube command by specifying how big you want your cuboid to be. The square brackets ([ ]) indicate a *vector* that organizes the three dimensions of your cuboid. The order of the numbers in the vector is important: 5 is the width of the cuboid along the x-axis, 10 is the length of the cuboid along the y-axis, and 20 is the height of the cuboid along the z-axis. Finally, mark the end of the statement with a semicolon (;).

Notice that one corner of the cuboid touches the *origin*: the point at which the three axes meet, represented by the coordinates (0, 0, 0).

Drawing Spheres with sphere

To draw a sphere, use the sphere command followed by the sphere's radius in parentheses to indicate its size. For example, the following statement draws a sphere with a radius of 10 units (Figure 1-3):

```
sphere(10);
```

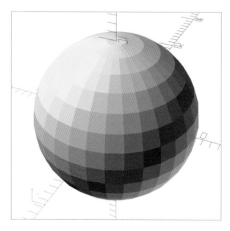

Figure 1-3: A sphere with a radius of 10 units

You can change the size of a sphere by changing its radius. Unlike cuboids, which might have three distinct measurements for width, length, and height, a sphere has the same measurements along all three axes. That's why the basic sphere command has only one number inside the parentheses. As with the cube command, mark the end of the code statement with a semicolon. But unlike with the cube command, OpenSCAD centers a sphere around the origin.

Drawing Cylinders and Cones with cylinder

To draw a cylinder, use the cylinder command followed by parentheses containing the cylinder's height and the length of the two radii of the circles that form its top and bottom. The following statement draws a cylinder with two radii of the same size (Figure 1-4):

```
cylinder(h=20, r1=5, r2=5);
```

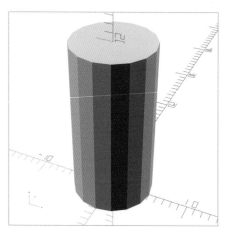

Figure 1-4: A cylinder with a height of 20 units, a bottom radius of 5 units, and a top radius of 5 units

Because keeping track of the cylinder's three parameters can be confusing, OpenSCAD allows you to label each parameter and include them in the command in any order. In parentheses, set the following values: h, which is the height of the cylinder along the z-axis; r1, which is the radius at the bottom of the cylinder; and r2, which is the radius at the top of the cylinder. As with the sphere and cube commands, use a semicolon to mark the end of the statement.

PARAMETER ORDER

It's perfectly fine to pass parameters to cylinder without labels for height and radii, so entering cylinder(15, 8, 8) is equivalent to cylinder(h=15, r1=8, r2=8). However, if you don't use labels, the parameters must be in the exact order for it to be read properly. If using labels, you can enter the parameters in any order, for example: cylinder(r1=8, r2=8, h=15).

The two radii of a cylinder don't need to have the same measurements. When they're different, the cylinder looks more like a cone with its top cut off (or, a truncated cone, according to mathematicians), as shown in Figure 1-5:

```
cylinder(h=20, r1=5, r2=3);
```

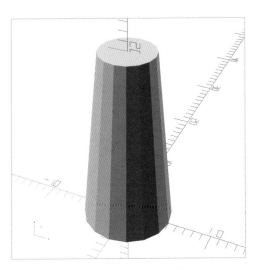

Figure 1-5: A cone with a height of 20 units, a bottom radius of 5 units, and a top radius of 3 units

You can draw a pointed cone, like the one in Figure 1-6, by assigning one of the radii a radius of 0:

```
cylinder(h=20, r1=0, r2=5);
```

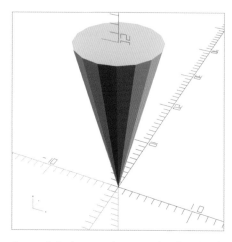

Figure 1-6: A pointed cone with a height of 20 units,
a bottom radius of 0 units, and a top radius of 5 units

Notice also that, unlike the shapes drawn with the sphere and cube commands, cylinders are centered around the z-axis, with one face touching the
xy-plane.

Importing 3D Models with import

OpenSCAD allows you to import shapes from other 3D design programs
if they're saved in the *.stl* format, which is a common format for 3D models. You can import these preexisting 3D shapes with the import command.
For example, use the following statement to import a popular file called
3DBenchy.stl (Figure 1-7):

```
import("3DBenchy.stl");
```

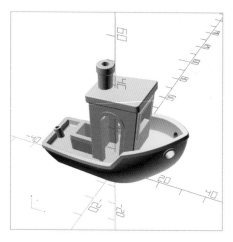

Figure 1-7: An imported 3D model of a boat often
used to calibrate 3D printers

To import a 3D shape, place the *.stl* file's name within parentheses after the import command. Enclose the name of the file in quotation marks (" ") to indicate that the filename is literal text and should not be interpreted by OpenSCAD. Note that you should save the *.stl* file in the same folder/directory as your OpenSCAD program, and be sure to save your OpenSCAD program before you generate a preview of your design; otherwise, OpenSCAD might have trouble finding the file. Mark the end of the statement with a semicolon.

Modifying Basic Shapes

Some of the basic ways to alter the shapes you draw with OpenSCAD include moving or smoothing them.

Moving Shapes

If the design you're creating has more than one shape, you'll need to know how to move those shapes around the Preview window. Otherwise, by default, they will sit on top of each other, and you may not be able to see the shapes of different sizes. For example, consider the following design (Figure 1-8):

```
cube([20, 10, 10]);
sphere(5);
cylinder(h=30, r1=2, r2=2);
```

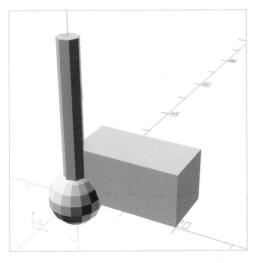

Figure 1-8: Multiple shapes drawn with default positioning

Centering Shapes with center=true

By default, the `sphere` command draws a sphere so that it's centered around the origin; the `cube`, `cylinder`, and `import` commands don't do this. If you want to draw other shapes so that they're also centered around the origin, add the `center=true` parameter inside the parentheses, as in this snippet (Figure 1-9):

```
cube([5, 10, 20], center=true);
```

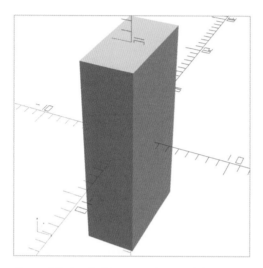

Figure 1-9: A cuboid centered around the origin

Now the cuboid's center will be at (0, 0, 0). You can also add the `center=true` parameter to cylinder shapes in order to center cylinders and cones around the origin. It's not possible to center imported shapes with `center=true`.

Moving Shapes to a Specific Location with translate

To move a shape to a specific location in the Preview window, use the `translate` operation. This operation modifies a shape as a whole so it's included right before the shape it's meant to modify.

For example, the following statement draws a cuboid that is shifted from its default position by 10 units in the negative direction along the x-axis, 20 units in the positive direction along the y-axis, and 0 units along the z-axis (Figure 1-10):

```
translate([-10, 20, 0]) cube([20, 10, 10]);
```

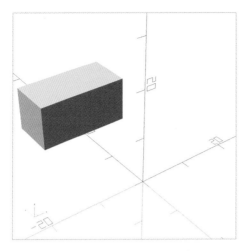

Figure 1-10: A translated cuboid with a starting
corner at (–10, 20, 0)

The translate operation uses square brackets to group the x, y, and z
dimensions into a vector. Similar to specifying the dimensions of a cube
shape, the order of the numbers in the vector is important. The first number
in the translation vector describes movement along the x-axis; the second
describes movement along the y-axis; and the third describes movement along
the z-axis. Finally, mark the end of the entire statement with a semicolon.

You may have noticed that the vector you use to modify the translate
operation moves the shape's starting corner—the corner that touches the
origin by default. Figure 1-11 shows how the translate operation moves the
cuboid relative to the origin (the original cube is shown in gray). You can
use the axes legend to predict the location of your shapes after the translate
operation has been applied.

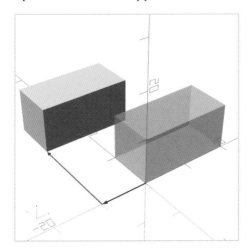

Figure 1-11: A cuboid moved 10 units along the
x-axis and 20 units along the y-axis, compared
with the same-sized cuboid at the origin

To create a more complex design, you may need to move shapes around in different configurations. Use the translate operation in front of a command to move it into a different position. For instance, the following statements draw a cuboid, a sphere, and a cylinder in one Preview window (Figure 1-12):

```
translate([-10, 10, 0]) cube([20, 10, 10]);
translate([20, 0, 0]) sphere(5);
translate([0, 0, -10]) cylinder(h=30, r1=2, r2=2);
```

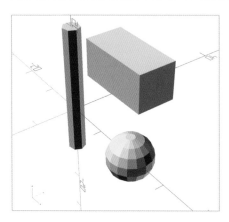

Figure 1-12: Three distinct shapes, translated from default positions

Both the sphere and cylinder move according to their respective center points, while the cube moves relative to the corner that touches the origin. Notice that the movement is different if you apply the same translation operations to a cube and cylinder that have been centered (Figure 1-13):

```
translate([-10, 10, 0]) cube([20, 10, 10], center=true);
translate([20, 0, 0]) sphere(5);
translate([0, 0, -10]) cylinder(h=30, r1=2, r2=2, center=true);
```

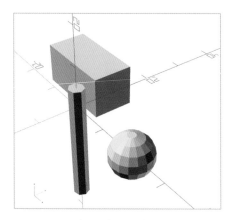

Figure 1-13: Three distinct shapes, translated from centered positions

Smoothing Curves with $fn

You might be wondering why the spheres and cylinders you've drawn so far don't appear to be round, but instead are formed by a series of flat panels. That's because OpenSCAD, like most 3D design software, uses a collection of straight lines to approximate a curve. To save on memory and reduce the processing time required to draw complex shapes, OpenSCAD uses a relatively small number of these lines by default. The cylinder shown in Figure 1-13, for example, uses only six line segments to approximate the curve of the circular faces of the cylinder.

To make your cylinders and spheres smoother, specify the number of line segments used to approximate a curve by including the $fn parameter. Setting $fn to 10, for instance, makes a cylinder look a bit rounder, because it draws the circumference of the cylinder with 10 line segments (Figure 1-14):

```
cylinder(h=20, r1=2, r2=2, $fn=10);
```

Figure 1-14: Approximating the curve of a cylinder
with 10 line segments

As with other parameters, include $fn in the parentheses within the command.

Although the cylinder in Figure 1-14 is rounder than a default cylinder, it's still not visibly round. Increase $fn to an even larger value in order to make the cylinder rounder (Figure 1-15):

```
cylinder(h=20, r1=2, r2=2, $fn=50);
```

With 50 line segments, the curve in this cylinder looks a lot smoother. After a certain point, though, increasing $fn will stop showing any visible effect. Also, note that OpenSCAD takes longer to generate shapes with large $fn values (as there are more details to generate), so be sure to consider the trade-off between smoothness and computational overhead when you set $fn. Generally, $fn=50 will produce a "roundness" that is more than sufficient.

Figure 1-15: A cylinder with a curve approximated with 50 line segments

Combining 3D Shapes with Boolean Operations

Sometimes you'll want to create shapes with features that are more complex than the basic shapes you've made so far. The *Boolean* operations in OpenSCAD allow you to combine multiple shapes, like cuboids, spheres, cylinders, and cones, into one shape (Figure 1-16). You can do this by using one of three operations: union, difference, or intersection.

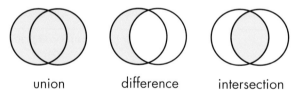

| union | difference | intersection |

Figure 1-16: An illustration of basic Boolean operations

The union operation groups two shapes together, the difference operation subtracts one shape from another, and the intersection operation keeps only the parts where two shapes intersect with each other.

Subtracting Shapes with difference

Let's start by subtracting shapes with the difference operation (Figure 1-17):

```
difference() {
    cube([10, 10, 10]);
    sphere(5);
}
```

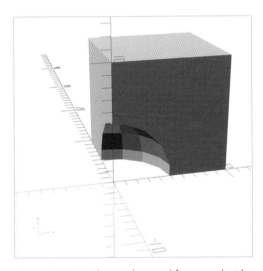

Figure 1-17: A sphere subtracted from a cuboid with the difference *operation*

Indicate a difference operation, followed by a set of parentheses, and then enter at least two commands between a set of curly brackets. Order matters when you use the difference operation; it keeps only the first shape, removing the parts of that shape where the remaining shapes intersect it. Notice in Figure 1-18 what happens when you exchange the order of the two shapes:

```
difference() {
    sphere(5);
    cube([10, 10, 10]);
}
```

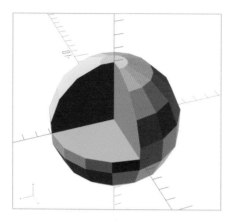

Figure 1-18: A cuboid subtracted from a sphere
with the difference operation

Reversing the operations creates a sphere with a slice missing, precisely where cube would have drawn a cuboid shape on top of the original sphere.

Debugging difference Operations with

It can be easy to lose track of the shape you're subtracting because it is no longer visible in the design. To make things easier, place a hash mark (#) in front of a subtracted shape to create a ghost version of the shape. The following code is identical to the code that drew Figure 1-17, except it uses a hash mark to render the sphere as a ghost-like image (Figure 1-19):

```
difference() {
    cube([10, 10, 10]);
    #sphere(5);
}
```

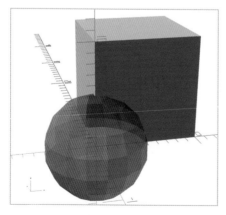

Figure 1-19: A ghost version of a subtracted
sphere to help with problem-solving

Use the hash mark to help you debug your designs, and then when your design is correct, be sure to remove the hash mark from your code.

Avoiding "Shimmering Walls" with the difference Operation

When subtracting shapes with the difference operation, you may sometimes end up with "shimmering walls" like those in Figure 1-20.

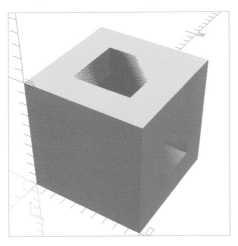

Figure 1-20: Two cuboids subtracted from a larger cuboid create shimmering walls

The shimmering walls appear because the subtracted shapes share a face with the shape they're being subtracted from. This creates an ambiguous scenario; should the face remain or be subtracted? Because of this concern, a model with shimmering walls isn't 3D-printable.

To solve this issue, only subtract shapes that extend slightly beyond the size of the outer shape (Figure 1-21).

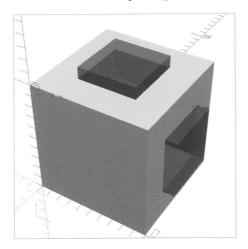

Figure 1-21: Two slightly larger cuboids subtracted from an outer cuboid

Once you've removed the ghost shapes, the remaining shape should contain no shimmering walls (Figure 1-22):

```
difference() {
    cube([10, 10, 10]);

    translate([-1, 2.5, 2.5]) cube([12, 5, 5]);
    translate([2.5, 2.5, -1]) cube([5, 5, 12]);
}
```

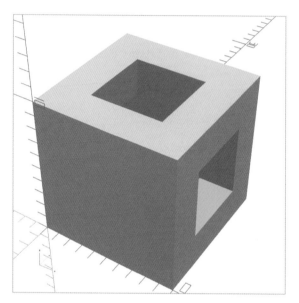

Figure 1-22: A subtracted shape that is fit for 3D printing

You should now be able to 3D-print this design.

Carving Out Overlapping Shapes with intersection

You can also carve away everything *except* the overlapping portion of two shapes by using the intersection operation (Figure 1-23):

```
intersection() {
    sphere(5);
    cube([10, 10, 10]);
}
```

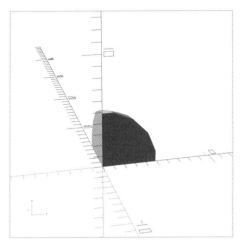

Figure 1-23: The cutout of an overlapping sphere
and cuboid, drawn with the intersection operation

First, indicate the intersection operation followed by parentheses, and then enter at least two commands between curly brackets. Unlike with the difference operation, the order in which you include the shapes doesn't matter with intersection.

Grouping Shapes with union

To group shapes into a single entity, use the union operation (Figure 1-24):

```
union() {
    cube([10, 10, 10]);
    sphere(5);
}
```

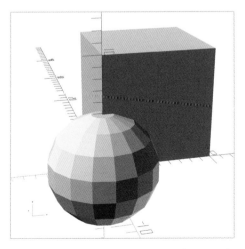

Figure 1-24: A sphere and a cuboid grouped
together with a union operation

The union operation combines all the shapes inside the curly brackets into one shape. Indenting all the lines that come between the curly brackets makes your code readable and easy to understand. Similar to intersection and difference, there's no way to modify the union operation, so you'll never need to put any information inside its parentheses.

Although it appears as if you can combine shapes by simply drawing them on top of each other, each shape will still remain a separate entity. This can be a problem when using the difference operation, as that operation subtracts only from the first shape inside the curly brackets. To avoid this problem, you can group multiple shapes into one shape by using the union operation. Include this grouped shape within difference as the first shape. For example, the following program uses the union operation to subtract a sphere from two shapes at once (Figure 1-25):

```
difference() {
    union() {
        cube([10, 10, 10]);
        cylinder(h=10, r1=2, r2=2);
    }
    sphere(5);
}
```

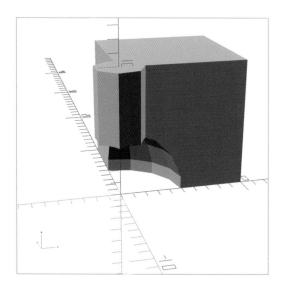

Figure 1-25: A sphere subtracted from a cylinder and a cuboid grouped together with union

OpenSCAD first combines the cube and cylinder into one shape, and then subtracts the sphere from that new shape. Without the union operation, OpenSCAD would, instead, subtract both the cylinder and sphere from the cuboid (Figure 1-26).

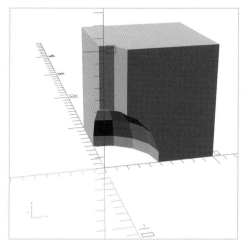

Figure 1-26: A sphere and a cylinder subtracted from a cuboid

Once you've created a complex shape with difference, intersection, or union, a computer can easily break it into geometric primitives to generate an accurate 3D model of your design. You can then print this complex 3D model on a 3D printer or import it into a 3D virtual reality program.

Getting Ready for 3D Printing

When you're ready to send your OpenSCAD design to another application for 3D printing, you'll need to export an *.stl* version of your design from OpenSCAD. You can then import this file into your 3D printing preparation software to adjust the settings, then turn it into a physical object with a 3D printer.

To export an *.stl* version of your design, first render your design by clicking the **Render** button (circled in red in Figure 1-27). Whereas Preview generates a quick picture of your model, Render fully calculates all of the surfaces needed to define the model. Especially complex designs require more surfaces and might have slow Render times as a result.

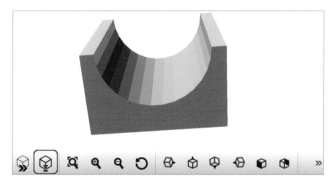

Figure 1-27: Rendering a design with the Render button

Finally, export your design as an *.stl* by selecting **File ▸ Export ▸ Export as STL** (Figure 1-28).

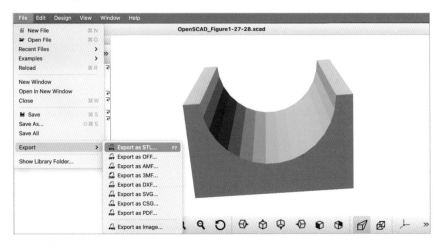

Figure 1-28: Exporting a design as an .stl file

Summary

Congratulations! You should now be able to create designs that include cuboids, spheres, and cylinders in any size and draw them in OpenSCAD's Preview window. You can also import 3D shapes, smooth curves, and move shapes to anywhere along the x-, y-, and z-axis. Finally, you also should know how to create complex designs out of basic shapes by grouping, subtracting, and cutting out overlapping shapes.

Here are some important points to remember:

- The name of an OpenSCAD command describes the type of shape you'd like to draw.
- Commands are followed by parentheses. Information inside parentheses ( ) modifies a command. The values inside the parentheses are called *parameters*. You can think of parameters as adjectives that describe characteristics of the shape.
- A semicolon (;) marks the end of most statements. Statements can include both commands and operations.
- Use the translate operation to move your shapes around the Preview window. Indicate the amount and direction of movement by changing the vector parameter of the translate operation.
- Square brackets ([ ]) collect numbers together to form a vector. The order of the numbers inside a vector is important.
- Boolean operations use curly brackets ({ }) to collect multiple shapes together. These curly brackets also form a complete OpenSCAD statement and do not require a semicolon to end the statement.
- Parentheses, square brackets, and curly brackets always come in pairs.

- $fn can be used as a parameter to change the smoothness of a single shape. You can also set $fn to a high value at the beginning of your code to generate smooth curves for every shape in a design. High values for $fn can result in slow rendering times.

- Use indentation to help make your code readable and easy to understand.

- A design must be *rendered* before it can be exported as an *.stl* file.

DESIGN TIME: 3D SHAPES

Practice your composition and design skills by building each of the complex shapes in Figure 1-29. We strongly recommend that you finish building each shape before moving on.

1. Mouse

2. Yo-yo

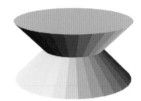

3. Spinner

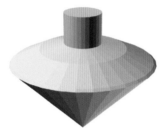

4. Epcot

5. Half-pipe

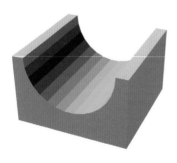

6. Ice cream cone

Figure 1-29: Practice drawing these shapes.

The following big projects will help you practice the commands covered in this chapter, and will introduce you to some basic considerations for using your 3D printer, such as printer resolution and temperature.

CALIBRATION PYRAMID

Building a calibration pyramid, shown in Figure 1-30, will help you determine whether you need to tweak the settings on your 3D printing preparation software. It will also help you practice using cube and translate.

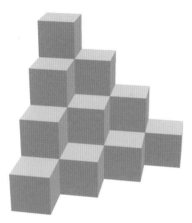

Figure 1-30: Calibration pyramid

- Try printing this at different *resolutions*. Try Low, Medium, and High quality. Compare print times and results.
- Try printing this pyramid at different sizes. Measure the pyramid after you print it. Check to make sure your physical measurements match the virtual measurements of your 3D model.
- Tweak your software settings so the pyramid has straight lines at all corners after it's printed.

(continued)

SMILEY-FACE PENDANT

To create the smiley-face pendant shown in Figure 1-31, you'll need to use your 3D printer to print large, flat shapes. Flat shapes can be difficult to print because they tend to curl.

Figure 1-31: Smiley-face pendant

If you have a heated bed on your 3D printer, use it. Vary the temperature of your heated bed to see which temperature works best for the type of filament you are using.

If you don't have a heated bed, inserting thin *helper disks* can keep prints from curling. Helper disks are 1 mm thick, so you can easily remove them after your print is complete. You can place helper disks around the perimeter of your design, and then easily cut them off after you've finished the print. Some 3D printing preparation software allow you to insert these discs automatically. Otherwise, you can insert ultra-thin cylinders in your OpenSCAD design.

HOLE-AND-PINS TEST

To design a hole-and-pins test, you'll use your 3D printer to print pieces that fit together, as shown in Figure 1-32. If you design this properly, the pins should fit snugly inside the holes.

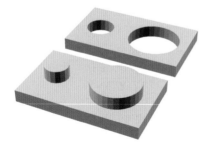

Figure 1-32: Hole-and-pins test

If you design the pins to be exactly the same size as the holes, the two pieces won't fit together. The pins should be slightly smaller than the holes. How much smaller depends on the type of filament you're using and your printer settings. Both the brand and type of plastic will make a difference.

2

MORE WAYS TO
TRANSFORM SHAPES

This chapter introduces a collection of
transformation operations that allow you
to have more control when creating complex
shapes. You'll learn how to rotate, reflect, and
scale shapes; combine them with a shared hull; and
round out their edges. These transformation operations will expand your modeling toolbox and allow
you to create even more complex designs.

OpenSCAD Shape Transformations

First, you'll learn how to use three transformation operations: rotate,
mirror, and resize. A *transformation operation* is a bit of code that comes
immediately before a shape to alter the shape's position, size, or

orientation. For illustrative purposes, we include a transparent gray outline in this chapter's examples to indicate where the original, untransformed shape would have appeared.

Rotating Shapes with rotate

By default, OpenSCAD draws shapes so they're oriented in a certain way. It draws sphere shapes centered at (0, 0, 0), for example, and cube shapes with a single corner at (0, 0, 0). Sometimes, though, you'll want your shape to have a different orientation.

One way to alter a shape's default position is to *rotate* it. To rotate a shape, specify the angle of rotation around each of the three axes, and express the angles of rotation in degrees, which can be positive or negative.

The following code snippet rotates a cuboid 90 degrees around the x-axis (Figure 2-1):

```
rotate([90, 0, 0]) cube([30, 20, 10]);
```

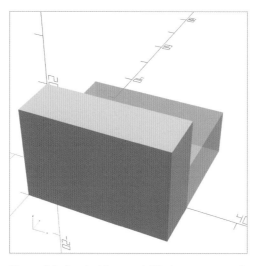

Figure 2-1: A cuboid rotated 90 degrees around the x-axis

First, write the name of the transformation, and then inside the parentheses, provide rotate with a vector in square brackets ([ ]) to group together the three axes of rotation. The first element in the vector is the degree of rotation around the x-axis, the second is the degree of rotation around the y-axis, and the third is the degree of rotation around the z-axis. Next, write the code for the shape you want to rotate. As always, use a semicolon (;) to end the entire statement.

Because you're rotating the shape 90 degrees around the x-axis, its position the x-axis stays fixed, and it gets a new position on the yz-plane.

The following code snippet rotates the same cuboid around the y-axis (Figure 2-2):

```
rotate([0, 180, 0]) cube([30, 20, 10]);
```

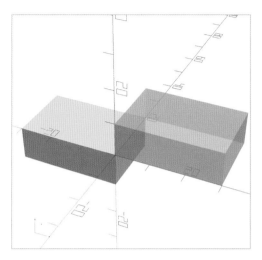

Figure 2-2: A cuboid rotated 180 degrees around the y-axis

In this case, the shape's position relative to the y-axis stays fixed, and its position on the xz-plane moves by 180 degrees.

You can also rotate a shape around two axes with a single operation, as in the following snippet (Figure 2-3):

```
rotate([-90, 0, -90]) cube([30, 20, 10]);
```

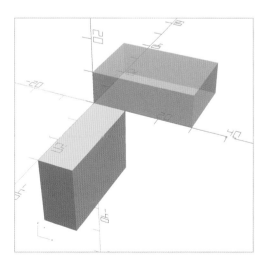

Figure 2-3: A cuboid rotated 90 degrees around the x-axis and 90 degrees around the z-axis

This cuboid is rotated around both the x- and z-axes. You might find it easier to imagine this operation as two separate transformations: one that rotates the shape around the x-axis and one that rotates it around the z-axis. To rotate the shape counterclockwise by 90 degrees in both directions, set the angle of rotation for those axes to –90.

Even though rotation around multiple axes is possible with the application of only one rotation operation, it's best to separate the various rotations into individual, repeated transformations. This is because it is sometimes hard to predict which rotation will be applied first. Consider the difference in the location of the cuboid when the rotation around the z-axis is applied before the rotation around the x-axis (Figure 2-4):

```
rotate([-90, 0, 0]) rotate([0, 0, -90]) cube([30, 20, 10]);
```

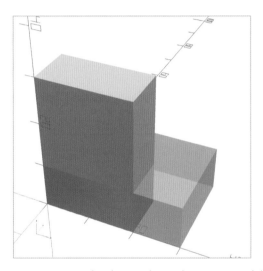

Figure 2-4: A cuboid rotated –90 degrees around the
z-axis, then rotated –90 degrees around the x-axis

Explicitly applying multiple rotations in their intended order will result in shapes ending up exactly where you'd like them to be after the rotations are applied.

Reflecting Shapes with mirror

Another way to change a shape's default position is to *reflect* it across an imaginary 2D plane with the mirror transformation. As you might expect from the name of the operation, mirror creates a mirror-like reflection of your shape. The following statement reflects a truncated cone across the yz-plane (Figure 2-5):

```
mirror([10, 0, 0])
  translate([0, 10, 0]) rotate([0, 90, 0]) cylinder(h=10, r1=5, r2=2);
```

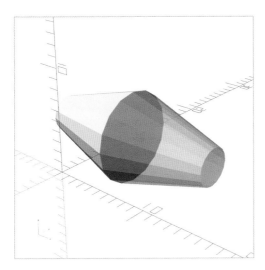

Figure 2-5: A truncated cone reflected across the yz-plane
via the vector [10, 0, 0]

The vector you pass to mirror contains the x, y, and z coordinates that define an imaginary point. OpenSCAD then draws an imaginary line from the origin to that point and uses the 2D plane that is perpendicular to that line at the origin as the *mirror*, or plane of reflection.

To clarify this, Figure 2-6 shows the "mirror" as a semitransparent plane.

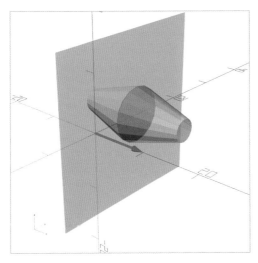

Figure 2-6: A truncated cone reflected across the yz-plane
via the vector [10, 0, 0]

The "mirror" is perpendicular to the vector, shown in green, drawn from (0, 0, 0) to (10, 0, 0). Notice that you don't have to use 10 as the x-axis value to create this mirror; any nonzero x-axis value would cause the mirror operation to behave the same way, as your goal is only to specify a vector that is perpendicular to the mirror. The *mirror plane* always contains the origin (0, 0, 0). In effect, the vector parameter of the `mirror` operation describes how the mirror is rotated.

The next statement reflects a cylinder across the xy-plane (Figure 2-7):

```
mirror([0, 0, 10]) cylinder(h=10, r1=2, r2=5);
```

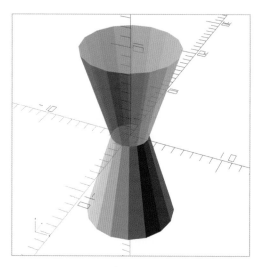

Figure 2-7: A cone reflected across the xy-plane via the vector [0, 0, 10]

This example defines a point at (0, 0, 10), and the line from the defined point to the origin is perpendicular to the xy-plane. The `mirror` operation is particularly useful for quickly creating complex shapes that involve symmetry. Using the `mirror` operation in such cases may save you time, as you can focus on designing only one half of the object, and then use `mirror` to create the second half.

Note that the `mirror` operation does not copy the shape; it moves the shape into the mirrored position. If you want a fully symmetrical shape, first create the shape, and then repeat it with the `mirror` operation in front of it.

Scaling Shapes with resize

The resize operation allows you to stretch or shrink specific dimensions of individual shapes. When you resize a shape, you can specify its exact dimension along each axis. By stretching a sphere across a single axis, for example, you can turn it into an ellipsoid (an elongated sphere).

The following code snippet uses resize to scale a sphere with a radius of 1 into an ellipsoid (Figure 2-8):

```
resize([10, 10, 20]) sphere(1, $fn=100);
```

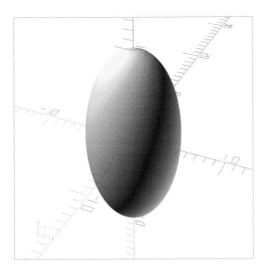

Figure 2-8: A sphere resized into an ellipsoid

Before writing the shape command, pass a vector to the resize operation to group together the new dimensions of the sphere along the x-, y-, and z-axes. As with all transformations, use a semicolon to end the entire statement.

The new ellipsoid stretches 5 units on either side of the origin along the x-axis, 5 units on either side of the origin along the y-axis, and 10 units on either side of the origin along the z-axis.

You could also use resize to transform a basic cylinder (Figure 2-9):

```
resize([10, 5, 20]) cylinder(h=5, r1=5, r2=5);
```

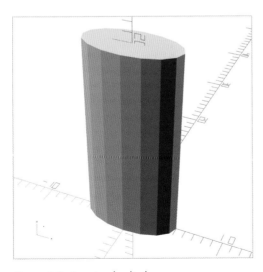

Figure 2-9: A resized cylinder

This statement resizes a basic cylinder with a height and two radii of 5 units so that the transformed cylinder stretches 10 units along the x-axis (through the origin), 5 units along the y-axis (also through the origin), and 20 units along the z-axis (from the origin).

More Ways to Combine 3D Shapes

In Chapter 1, you learned about three Boolean operations that allow you to combine multiple 3D shapes: union, difference, and intersection. You can also combine two shapes into one with the hull and minkowski operations.

Combining Shapes with hull

The hull operation creates a convex *hull* (or skin) around two shapes. To understand this, imagine stretching a balloon tightly around two or more shapes in order to create a single shape. For example, the following code creates a balloon surrounding both a sphere and a cube (Figure 2-10):

```
hull() {
    translate([10, 0, 0]) sphere(8);
    translate([-10, 0, 0]) cube([4, 4, 4], center=true);
}
```

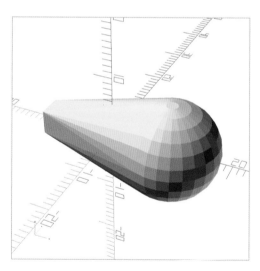

Figure 2-10: A hull stretched around a small cube and a big sphere

The hull operation has the same syntax as the Boolean operations described in Chapter 1. It can combine two or more shapes, and as with the union operation, the order of shapes does not matter.

Combining Shapes with minkowski

The minkowski operation creates a *Minkowski sum* of a collection of shapes. This means it wraps the edges of one shape with the characteristic of a second shape. The following example wraps a sphere around the edges of a cylinder to create rounded edges (Figure 2-11):

```
$fn=50;
minkowski() {
    cylinder(h=15, r1=5, r2=5);
    sphere(4);
}
```

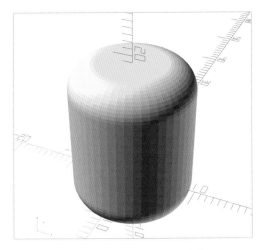

Figure 2-11: A sphere used to smooth the corners of a cylinder

The minkowski operation also has the same syntax as the Boolean operations described in Chapter 1. In this example, the edges of the cylinder become rounded edges because the smaller sphere has been embossed along the edges of the cylinder. It's important to note that the minkowski operation produces a larger shape than the original cylinder, because wrapping the sphere around the original cylinder adds volume.

Combining Transformations

You can combine transformation operations by writing one operation in front of another. For example, the following code snippet applies the rotate operation before translate on each of three cylinders (Figure 2-12):

```
translate([5, 0, 0]) rotate([90, 0, 0]) cylinder(h=10, r1=4, r2=4);
translate([5, 0, 0]) rotate([0, 90, 0]) cylinder(h=10, r1=4, r2=4);
translate([5, 0, 0]) rotate([0, 0, 90]) cylinder(h=10, r1=4, r2=4);
```

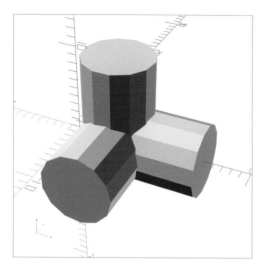

Figure 2-12: Three cylinders, rotated and then translated

OpenSCAD first executes the innermost transformation (the operation directly to the left of a shape), then applies the outermost transformation. If you applied the transformations in the reverse order, you'd get a different result. The next snippet applies the translate operation before the rotate operation (Figure 2-13):

```
rotate([90, 0, 0]) translate([5, 0, 0]) cylinder(h=10, r1=4, r2=4);
rotate([0, 90, 0]) translate([5, 0, 0]) cylinder(h=10, r1=4, r2=4);
rotate([0, 0, 90]) translate([5, 0, 0]) cylinder(h=10, r1=4, r2=4);
```

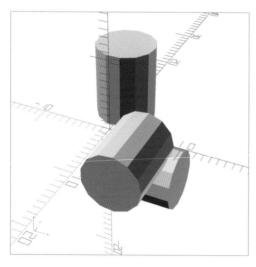

Figure 2-13: Three cylinders, translated and then rotated

You get different results because OpenSCAD applies operations in order, starting with the transformation operation closest to the shape.

Summary

This chapter introduced several important operations for transforming shapes. You can now move, rotate, reflect, and resize shapes. You can also combine two shapes by forming a hull around them or by smoothing the corners of one shape with another.

Here are some important points to remember:

- You can apply transformation operations to single shapes and combined shapes.
- Combining shapes with the `union` operation can reduce the number of transformation operations that you need to apply to a complex design.
- Applying a series of `rotate` operations is often easier to manage than combining rotations into one `rotate` operation.
- Reflecting combined shapes with `mirror` can save you time when you're building symmetrical designs.
- When you're applying multiple transformation operations, order matters.
- The transformation operation closest to the shape is applied first.

In the next chapter, you'll learn how to convert 2D shapes into 3D shapes, apply transformation operations to 2D shapes, and create surprisingly complex 3D shapes by combining and operating on basic 2D shapes.

Before moving on to Chapter 3, practice the skills you learned in this chapter by building each of these complex shapes (Figure 2-14).

1. Heart

2. OpenSCAD logo

3. Guitar pick

4. Snowman

5. Modern table

6. Top hat

Figure 2-14: Practice building each of these shapes.

Hone your transformation skills with these bigger, more difficult projects.

GAME DIE

This project will help you when you're missing a die in the *Monopoly* box. Build your own die (Figure 2-15) to practice the operations you learned in this chapter.

Figure 2-15: Game die

3D-Printing Tip for the Game Die

The interior of a 3D-printed object consists of two parts: the infill and the shell. The *infill* is the interior volume of the object. The *shell* is the thick wall that forms the exterior shape of the design.

The *fill density* of a 3D-printed object describes how much volume of the object's interior will be filled with plastic. It's usually laid out in a crosshatched pattern to save time and plastic (although it's certainly possible to change this pattern). Try varying the fill density of your die to see how lower fill densities decrease printing times, while higher fill densities increase them.

Since your die won't be undergoing much structural stress, a 5 to 10 percent fill density should optimize both time and materials.

(continued)

DESKTOP ORGANIZER

When you're done playing with your die, put your design skills to a more practical use by building this desktop organizer (Figure 2-16) to hold your pencils and paper clips.

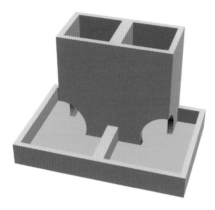

Figure 2-16: Desktop organizer

3D-Printing Tip for the Desktop Organizer

Shell thickness is the thickness of the outer wall of the 3D-printed design. It describes the number of layers on the outside of the print. A thicker outer wall makes your object much stronger, which is important if your object is going to undergo repeated use and stress. Increasing the shell thickness of your print can be a good way to increase the durability of your 3D-printed object without having to dedicate more resources to increasing the fill density.

Shell thickness is often described in terms of *nozzle diameters* (as in, the size of the hole that the melted plastic squeezes though in your 3D printer). The default shell thickness is often two nozzle diameters, which is about 0.8 mm. Try varying the shell thickness of this design to see the effect it has on print time and material usage.

3

2D SHAPES

You're now familiar with a good collection of basic OpenSCAD instructions for modeling simple 3D shapes, and you've seen operations that can transform those basic shapes into more complex designs. This chapter will teach you how to create and combine 2D shapes in order to build even more sophisticated 3D designs.

We'll start by showing you how to draw basic 2D shapes, and then we'll describe how to build on those basic 2D shapes to create elaborate 3D designs. Using 2D shapes will allow you to create designs that are not possible to build with the 3D shapes and operations you've learned so far. In addition, knowing how to create 2D shapes is useful when you're designing for other digital fabrication techniques, such as laser cutting, though that's beyond the scope of this book.

Drawing Basic 2D Shapes

As with 3D shapes, you can build complex 2D shapes based on a few built-in 2D primitives, called circle, square, and polygon.

Drawing Circles with circle

The circle command allows you to draw a 2D circle by specifying its radius, like the sphere command from Chapter 1. For example, the following statement draws a circle with a radius of 20 units (Figure 3-1):

```
circle(20);
```

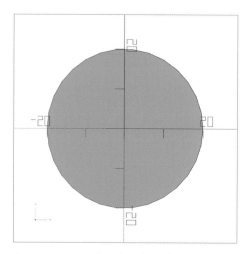

Figure 3-1: A rendered circle with a radius of 20 units

Clicking the **Preview** button renders your circle with a slight depth (Figure 3-2).

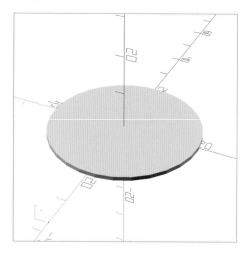

Figure 3-2: A previewed circle with a radius of 20 units

However, 2D shapes have no depth. They exist only in the xy-plane. To see 2D shapes in their true form, without depth, use the **Render** button. (Note that it's not possible to mix 2D and 3D shapes in Render mode.) Because 2D shapes have no depth, it's often easiest to create 2D designs by using the Top-view icon on the toolbar (Figure 3-3).

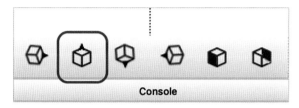

Figure 3-3: Top-view icon

Drawing Rectangles with square

The 2D square command, which draws rectangles, specifies x and y dimensions as a single vector parameter. The following statement draws a rectangle that extends 25 units along the x-axis and 10 units along the y-axis (Figure 3-4):

```
square([25, 10]);
```

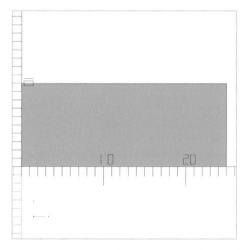

Figure 3-4: A rectangle with a width of 25 and height of 10 units

Use the square command to indicate that you want to draw a rectangle, followed by a set of parentheses. Within the parentheses, put square brackets, and then within those, enter the dimensions of the square, separated by a comma. This 2D vector requires only x and y dimensions, as opposed to the 3D vector (x, y, and z) required by the 3D cube shape. The first number in the vector represents the width of the square along the x-axis. The second number in the vector represents the length of the square along the y-axis.

Remember that you'll need to click the **Render** button to see the rectangle as a 2D shape.

Drawing Polygons with polygon

If you want to create a 2D shape that isn't built into OpenSCAD, you can create your own 2D shapes with the polygon command.

The following statement uses the polygon command to draw a triangle with vertices at [0, 0], [10, 0], and [10, 10] (Figure 3-5):

```
polygon([ [0, 0], [10, 0], [10, 10] ]);
```

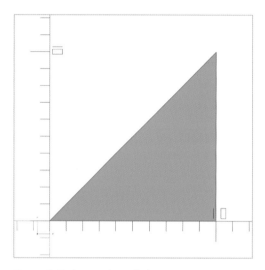

Figure 3-5: A triangle with three vertices

A polygon is defined by a list of the shape's corners, called *vertices*. Each vertex in this list is a vector containing the coordinates of a corner point in the polygon. Group each vertex as a vector within square brackets, then add an extra set of brackets around the entire list of vertices to organize the collection as a vector of vectors.

Be sure to list the vertices in order, as though you were walking around the edge of the polygon (in either direction). Also, you don't need to specify the starting point twice; OpenSCAD will finish the polygon for you automatically.

Since polygons can have any number of vertices, you can create increasingly complex shapes, like this one with eight vertices drawn with the following statement (Figure 3-6):

```
polygon([
  [ 0,  0], [20,  0],
  [20,  5], [ 5,  5],
  [ 5, 10], [20, 10],
  [20, 15], [ 0, 15]
]);
```

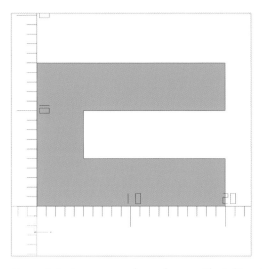

Figure 3-6: A more complex polygon with eight vertices

Drawing Words with text

Another way to use 2D shapes in your designs is to create symbolic patterns, such as words. Using textual elements in your designs can be useful for personalization. You may also want to use emoji fonts to access pre-drawn symbols or simply stamp a version or serial number onto your design.

Use the text command to draw text shapes in OpenSCAD. Text in OpenSCAD (as in other programming languages) is considered a *string of characters*. Since a string of characters can be arbitrarily long, quotation marks (" ") are used to indicate the beginning and end of the text string. Text strings can contain letters, punctuation, numbers, and (if the font used supports Unicode) emoji characters.

This statement creates the string "Hello, OpenSCAD" (Figure 3-7):

```
text("Hello, OpenSCAD", size=10);
```

Figure 3-7: Creating a 2D text shape

Follow the text command with parentheses containing a string of characters. The strings should start and stop with double quotes (" "). The parentheses can also contain an optional size parameter, which sets the text size to 10 in this case. Notice in Figure 3-7 that the tallest letters in the string reach the first tick mark (which represents 10 units) on the y-axis.

The size parameter is optional for text shapes. If you leave off the size parameter, the default text size is 10. Another optional parameter for drawing text shapes is font. You can also use the optional font parameter to draw text in any font installed on your computer. The following statement draws a string of text in Courier font (Figure 3-8):

```
text("Hello, OpenSCAD", font="Courier");
```

Figure 3-8: Changing the text shape's font to Courier

NOTE   *If you don't know the names of the fonts installed on your computer, you can ask OpenSCAD to give you a list by selecting* **Help ▶ Font List** *from the menu.*

Fonts that support Unicode characters will often contain emoji. You can draw any character supported by the font, including emoji shapes (Figure 3-9):

```
text("♕", font="Arial Unicode MS");
```

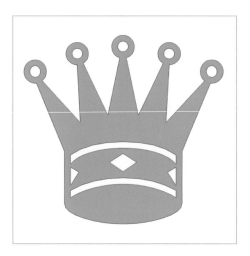

Figure 3-9: Using text to draw a crown emoji

It's also possible to draw numeric values with the text command. If you want to create a shape with a numeric value (Figure 3-10), be sure to convert the value to a string with the str function:

```
text(str(123), size=20);
```

Figure 3-10: Drawing a text shape with numbers

Rather than putting the number between quotation marks, apply the str function to a numeric value in order to turn it into a string. This is particularly helpful when the numeric value is stored in a variable, as we'll see in Chapter 4.

Applying Transformation and Boolean Operations on 2D Shapes

You can apply the same transformation and Boolean operations you learned in Chapters 1 and 2 to 2D shapes—and it's done pretty much the same way as when you apply them to 3D shapes. The only difference is that instead of requiring 3D vectors, the translate, mirror, and resize operations require 2D vectors containing x- and y-coordinates, and the rotate operation requires only a single angle of rotation (for the z-axis).

For example, the following design uses translate, difference, and rotate to draw an askew rectangle with three circles cut out of it (Figure 3-11):

```
rotate(30) {
    difference() {
        square([120, 40]);
        translate([20, 20]) circle(15);
        translate([60, 20]) circle(15);
        translate([100, 20]) circle(15);
    }
}
```

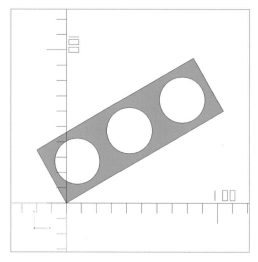

Figure 3-11: Transformation and Boolean operations on 2D shapes

Just as with the 3D shapes, the order in which you apply transformations and Boolean operations on a 2D shape will affect the arrangement and placement of the resulting shape. Consider the difference between subtracting a circle from a square versus subtracting a square from a circle. The following `difference` operation subtracts a circle from a square (Figure 3-12):

```
difference() {
    square([5, 5]);
    circle(5, $fn=50);
}
```

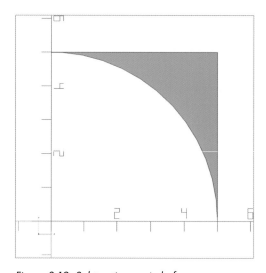

Figure 3-12: Subtracting a circle from a square

And this `difference` operation subtracts a square from a circle (Figure 3-13):

```
difference() {
    circle(5, $fn=50);
    square([5, 5]);
}
```

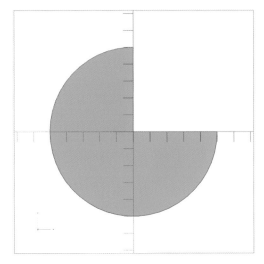

Figure 3-13: Subtracting a square from a circle

Extruding Shapes Vertically with linear_extrude

You can't 3D-print 2D shapes directly, but you can use them as building blocks for creating 3D shapes (which can then be 3D-printed as physical objects). This section describes two of OpenSCAD's powerful operations for creating 3D shapes from 2D shapes.

The `linear_extrude` operation takes a flat shape and "lifts" it up along the z-axis while building walls corresponding to the shape's initial boundary. The following statement extrudes the letter *A* into a 3D shape with a height of 5 units (Figure 3-14):

```
linear_extrude(5) text("A");
```

The `linear_extrude` operation takes a single parameter, the height of the 3D shape you're creating, followed by the 2D shape you'd like to stretch into 3D. As with the transformation operations you already know, end the entire statement with a semicolon.

You could also provide the `linear_extrude` operation the optional parameters of `twist`, `slices`, and `scale` to build more complex 3D shapes. The `twist` parameter specifies an angle at which to twist the 2D shape during extrusion. The `slices` parameter controls how smooth a twist will be—specifically, how many segments will be used to complete the twist.

Since extrusion extends a shape upward, each of these segments will turn into a horizontal "slice," which is why the parameter is named slices. If you don't specify it, OpenSCAD will choose a relatively coarse value. The scale parameter changes the size of the 2D shape during extrusion.

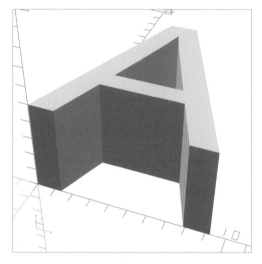

Figure 3-14: Linear extrusion of a 2D shape into a 3D shape

Use all of these parameters to transform a rectangle into the 3D shape drawn in Figure 3-15:

```
linear_extrude(100, twist=30, slices=25, scale=1/3) {
  square(100, center=true);
}
```

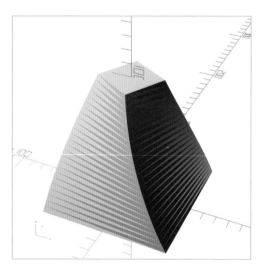

Figure 3-15: Twisting, scaling, and extending a 2D shape into a 3D shape with 25 horizontal slices

The parameters twist, slices, and scale are optional. Although this example shows all three parameters used at once, you can use any variation, such as only scale or only twist.

Extruding Shapes Along a Circle with rotate_extrude

Rather than extruding a 2D shape along a linear path, use the rotate_extrude operation to move the 2D shape along a circular path, which creates a donut-like shape called a *torus* (Figure 3-16):

```
rotate_extrude() {
  translate([100, 0]) circle(40);
}
```

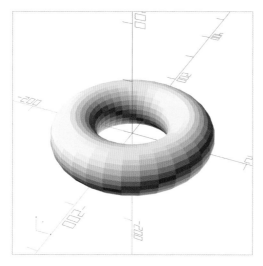

Figure 3-16: The rotate_extrude operation of a 2D circle into a 3D torus

The rotate_extrude operation is a two-step process that first rotates the 2D shape by 90 degrees around the x-axis, then moves the 2D shape in a circle around the z-axis. If you were to cut out a slice of the resulting donut, the shape of that slice would look like the original 2D shape.

When using rotate_extrude, take care to ensure that the shape doesn't rotate into itself. In the code that draws Figure 3-16, you do this by first translating the shape away from the z-axis so that no parts of the 2D shape are touching the z-axis.

The rotate_extrude operation also takes an optional angle parameter that allows you to specify the angle of rotation. Figure 3-17 demonstrates a circle that has been extruded along a 135-degree rotation around the z-axis.

```
rotate_extrude(angle=135) {
  translate([100, 0]) circle(40);
}
```

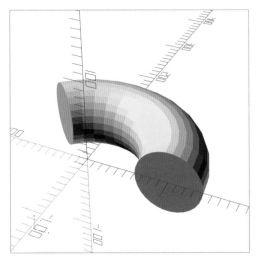

Figure 3-17: The rotate_extrude *with a 135-degree angle parameter*

TIPS FOR USING MULTIPLE LINES INSTEAD OF SINGLE LINES

Though we use curly brackets to help visually organize these rotate_extrude examples, curly brackets are optional if they enclose a single shape. So this multiline statement

```
rotate_extrude() {
    translate([100, 0]) {
        circle(40);
    }
}
```

is the same as this single-line statement:

```
rotate_extrude() translate([100, 0]) circle(40);
```

And, though there are no curly brackets in the multiline statement

```
polygon([
  [ 0,  0], [20,  0],
  [20,  5], [ 5,  5],
  [ 5, 10], [20, 10],
  [20, 15], [ 0, 15]
]);
```

we used for Figure 3-6, we could rewrite this statement as

```
polygon([[ 0,  0], [20,  0], [20,  5], [ 5,  5], [ 5, 10], [20, 10],
[20, 15], [ 0, 15]]);
```

or

```
polygon([[0,0],[20,0],[20,5],[5,5],[5,10],[20,10],[20,15],[0,15]]);
```

OpenSCAD ignores both the spaces between elements and new lines, so you have some flexibility in how you organize your code to make it more (or less) readable. While using indentation, new lines, and curly brackets can help communicate the nuances of a complex sequence of operations, consolidating elements onto one line can also be useful once you become more comfortable coding with OpenSCAD.

Growing and Shrinking a Shape with offset

Imagine you want to build a fancy cross-shaped cookie cutter. You now know how to create a cross shape by performing a union of two rectangles, and you know how to extrude it by using linear_extrude to make it 3D. But to specify the wall thickness, you need the offset operation, which allows you either to grow or shrink a shape by a specific amount. Use offset to hollow out your cookie cutter by shrinking one cross, and then subtract the small cross from the larger one.

In the following design, pass offset a negative value to shrink your 2D cross (Figure 3-18):

```
offset(-2) {
  union() {
    square([100, 30], center=true);
    square([30, 100], center=true);
  }
}
```

Place the code for the 2D shapes to offset in curly brackets following the offset operation. In parentheses, specify the amount (in millimeters) to offset. A positive value will grow a shape, and a negative value will shrink a shape.

NOTE   *When shrinking a shape, the inner corners become rounded, but when growing a shape, the outer corners become rounded. Experiment with this to build intuition on how offset works.*

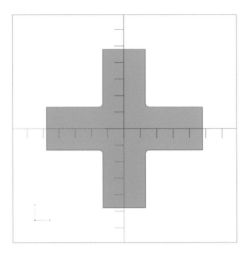

Figure 3-18: Shrinking an object by passing
offset a negative value

Now you can reuse that code to build the walls of your cross-shaped
cookie cutter (Figure 3-19):

```
linear_extrude(30) {
❶ difference() {
  ❷ union() {
      square([100, 30], center=true);
      square([30, 100], center=true);
    }
  ❸ offset(-2) {
      square([100, 30], center=true);
      square([30, 100], center=true);
    }
  }
}
```

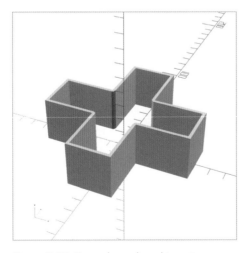

Figure 3-19: Cross-shaped cookie cutter

Define two squares to create the outer cross with a union operation ❷. Next, define two more squares to create the inner cross ❸, shrink that cross with offset, and then subtract it from the outer cross ❶. This leaves you with a hollowed-out cross shape.

Importing 2D Shapes with import

Just as with 3D shapes, you can import 2D shapes from files created in other 2D design programs. OpenSCAD supports importing the *.dxf* and *.svg* 2D file formats. These formats are commonly used with popular 2D vector graphic design tools, such as Adobe Illustrator and Inkscape (an open source alternative to Adobe Illustrator). OpenSCAD only supports importing shapes that are closed polygons, containing no "open-ended" sections. Also, make sure you convert all segments in a *.dxf* file to straight lines.

The syntax of the import command is the same for importing both 2D and 3D shapes. You just need to pass the filename in quotation marks to import, and make sure the file is saved in the same folder/directory as your project. For example, use the following statement to import the drawing in Figure 3-20:

```
import("drawing.dxf");
```

Figure 3-20: An imported .dxf vector graphic

Even though the imported file looks round, it actually consists of many short line segments, similar to the polygons you learned to create earlier in this chapter. Inkscape was used to draw this 2D smiley-face shape. An important final step in the process was to convert all of the line segments in the shape to very small straight lines.

Once you import a 2D shape, it behaves exactly like a built-in shape, and you can transform it and combine it with other shapes. The following statement first imports the smiley face shown in Figure 3-20, then extrudes it into the shape shown in Figure 3-21:

```
linear_extrude(height=500, scale=3) import("drawing.dxf");
```

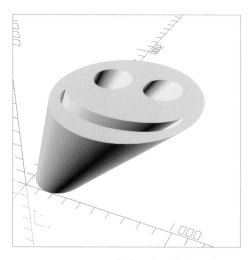

Figure 3-21: An extruded and scaled .dxf vector graphic

Now you have a 3D smiley-face shape that you can 3D-print as a stamp.

Summary

In this chapter, you learned how to design and create 3D shapes based on 2D shapes. You now should be able to create, combine, and transform simple 2D shapes like circles, rectangles, polygons, and text. You can create both internal and external outlines of 2D shapes with the offset operation, import vector graphics, and transform 2D shapes into 3D shapes.

By now you should be able to imagine a wide variety of designs that you could create with OpenSCAD 2D and 3D shapes. Sometimes it's easier to build a complex 3D design by thinking about its 2D shadow first, and then you can stretch the 2D shadow into 3D.

Here are some important points to remember when working with 2D shapes:

- Rendering a 2D design will display the actual 2D view of the shape, while a Preview window of the design will appear to add a small amount of height along the z-axis.

- 3D shape transformation vectors require three parameters: [x, y, z]; most 2D shape transformation vectors require only two parameters: [x, y].

- 2D rotations need only a single parameter: a number to represent the angle of rotation within the xy-plane.
- The Top view will often give you the best perspective when designing your 2D shapes.
- Extruding 2D shapes and text is necessary in order to combine them with 3D shapes.
- Text strings start and stop with double quotes.
- You can use the text shape to draw numeric values by converting the value to a string with the str function.
- Only fonts that support Unicode can be used to draw emoji, but think of how much fun you could have extruding emoji shapes!
- No part of a 2D shape can cross the z-axis when you use rotate_extrude on that shape.
- Think of 2D shapes as a "cross section" of the resulting 3D shape from a rotate_extrude operation.

Before moving on to Chapter 4, practice the skills you learned in this chapter by building each of the complex designs in Figure 3-22.

1. Fruit bowl

2. House

3. Stamp

4. Space Needle

5. Nice day

6. Star

Figure 3-22: Practice drawing these 2D designs.

Continue to practice the skills you learned in Chapters 1 through 3 with these three big projects.

STORYTELLING DICE

Use the basic game die you created in Chapter 2 to generate a collection of storytelling dice (Figure 3-23). Create dice for nouns, verbs, decisions, animals, heroes, villains, or any collection of related concepts. This will help you practice using and placing text shapes.

Figure 3-23: Storytelling dice

3D-Printing Tip for the Storytelling Dice

OpenSCAD is "unit-less." Often, but not always, if you import an *.stl* file to prepare it for 3D printing, the software will use millimeters as the unit for your design. It's important to adjust final dimensions for your 3D model as the last step before printing your model. Use the scaling features of your 3D-printing preparation software to check/change the final dimensions of your 3D print.

Play around with resizing your storytelling dice. What size makes the most sense? Be sure to make the dice big enough so you can read the text on all sides.

(continued)

PROJECT BOX FOR STORYTELLING DICE

Create a project box to hold your storytelling dice (Figure 3-24). Practice using the offset opera-tion to create a ridge in the wall of the box to keep the lid firmly in place. Don't forget to make small adjustments to the measurements of the lid so that the inner ridge fits snugly inside the box.

Figure 3-24: Project box

3D-Printing Tip for the Project Box

This box needs to be big enough to hold your storytelling dice with the lid in place. Most 3D-printing preparation software will allow you to change the size of your 3D model either by specifying a scalar percentage for each axis or by setting an absolute size for a certain dimension. You can set the other axes and dimensions to scale either uniformly or not at all.

TROPHY

Create a trophy (Figure 3-25) by using the shapes and operations introduced in this chapter. Start by designing a 2D profile of the cup and stem that you can rotate around the z-axis with rotate_extrude. Notice the embellishments on some of the edges of the trophy.

Figure 3-25: Trophy

3D-Printing Tip for the Trophy

Each 3D printer has its own *build volume* that determines the maximum measurement for each dimension of your 3D print. Be sure to stay within the build dimensions for your printer when you resize your model. In fact, your 3D-printing preparation software will warn you if you're exceeding your printer's dimensions.

Try to print this trophy at a larger size than the build volume of your 3D printer. You can accomplish this by splitting your final trophy into two parts: the base and the trophy. By using the difference operation, you can split your model into two pieces. Export each of these as separate *.stl* files, each containing a different part of the trophy. Then, scale each part in the 3D-printing preparation software so that you can print each separate piece to be as large as possible. A little superglue will allow you to recombine the two pieces into one extra-large trophy.

4

USING LOOPS AND VARIABLES

Starting with this chapter, you'll learn ways to use OpenSCAD to work smarter, not harder. First, you'll learn to use a very useful programming tool called a *loop*. Loops let you draw many similar shapes with only a few lines of code.

This is particularly useful when your designs have repeated features. For instance, if you're creating a model of the Empire State Building, typing one individual statement for each window in the building would consume a lot of time. With a loop, you can repeat a single window along a fixed pattern so OpenSCAD can take care of the tedious work of copying and pasting the same window many times. You'll also learn how to use variables to keep track of important data related to your designs. Because these new OpenSCAD tools will allow you to create more complicated designs, you'll also learn how to use comments to leave notes for yourself and other collaborators on your design.

Leaving Notes with Comments

In this chapter, the designs are a bit more complex than in previous chapters, so we'll use comments in the coding examples to explain important details in our designs. *Comments* provide a way for you to leave notes to yourself and others who might read your code. OpenSCAD ignores comment statements, as they are meant only as notes for the humans who read them rather than as instructions for OpenSCAD to draw a particular shape.

Writing Single-Line Comments with //

Single-line comments start with // and continue until the end of the line. They are useful for leaving short notes so you can remember later what your thought process was when you were creating your OpenSCAD design.

Writing Multiline Comments with /* */

Multiline comments begin with /* and end with */. OpenSCAD ignores everything inside a multiline comment. Multiline comments are useful for temporarily ignoring parts of your design when you want to focus on a particular element. Multiline comments make it easy to ignore multiple statements at once.

The following code shows single-line and multiline comments, which results in exactly one shape being drawn (a cuboid, Figure 4-1), as the other OpenSCAD statements are enclosed in comments and ignored:

```
cube([5, 10, 20]);

//sphere(5);

/*
cylinder(h=5, r1=10, r2=10);
cube([50, 50, 50]);
*/
```

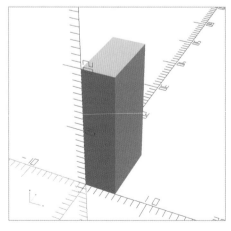

Figure 4-1: A single cube among a collection of comments

Repeating Code with a for Loop

The main focus of this chapter is on getting OpenSCAD to take care of the tedious and error prone "copy-and-paste" approach to typing very similar statements in order to draw a collection of similar shapes. If, for example, you want to draw 10 identical cylinders on a straight line, you could write 10 statements—one for each cylinder—changing only the vector parameter in the translate operation to prevent overlap, as in the following design (Figure 4-2):

```
translate([10, 30, 0]) cylinder(h=4, r1=4, r2=4);
translate([20, 30, 0]) cylinder(h=4, r1=4, r2=4);
translate([30, 30, 0]) cylinder(h=4, r1=4, r2=4);
translate([40, 30, 0]) cylinder(h=4, r1=4, r2=4);
translate([50, 30, 0]) cylinder(h=4, r1=4, r2=4);
translate([60, 30, 0]) cylinder(h=4, r1=4, r2=4);
translate([70, 30, 0]) cylinder(h=4, r1=4, r2=4);
translate([80, 30, 0]) cylinder(h=4, r1=4, r2=4);
translate([90, 30, 0]) cylinder(h=4, r1=4, r2=4);
translate([100, 30, 0]) cylinder(h=4, r1=4, r2=4);
```

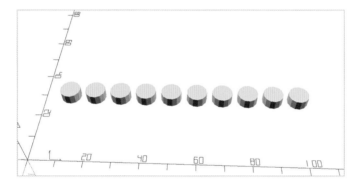

Figure 4-2: A row of cylinders drawn with 10 separate statements or drawn with a single for loop

Notice that the only change from one cylinder to the next is the increased position along the x-axis: the x position of the first cylinder is 10, the x position of the second cylinder is 20, and so on, until the last cylinder is drawn at an x position of 100.

Instead of using 10 separate statements, you can use a single for loop to generate this collection of cylinders. You just need to write a loop that draws the first cylinder 10 units from the x-axis, then increases the x position by 10 units every time a new cylinder is drawn, until drawing the last cylinder 100 units from the axis.

The following pseudocode shows the for loop syntax:

```
for (variable = [start: increment: end]) {
  // one or more statements to be repeated
}
```

The for keyword indicates that you want to repeat OpenSCAD statements. Then you create a *variable* to keep track of the changing value after each repetition. The *variable* has a *start* value, an *increment* value, and an *end* value. Similar to grouping multiple shapes together in order to apply a single transformation, use curly brackets ({ }) to enclose all of the statements you want to repeat.

The following example uses a single for loop to draw 10 cylinders instead of using 10 separate statements:

```
for (❶x_position = [10:10:100]) {
    translate([x_position, 30, 0]) cylinder(h=4, r1=4, r2=4);
}
```

A variable called x_position ❶ keeps track of the position of each cylinder. This variable has an initial value of 10; then every time the for loop repeats, the value of x_position increases by 10 so that the next cylinder is drawn 10 units farther along the x-axis. Once x_position is equal to 100, the last cylinder is drawn and the loop stops repeating. The resulting drawing will look the same as using 10 separate statements, as shown in Figure 4-2.

You can use loops to repeat shapes along many types of patterns. Figure 4-3 shows a cone repeating in a rotational pattern around the z-axis, and here's the corresponding for loop:

```
for (angle=[0:45:315]) {
    ❶rotate([0, 0, angle]) ❷translate([10, 0, 0]) ❸cylinder(h=5, r1=2, r2=0);
}
```

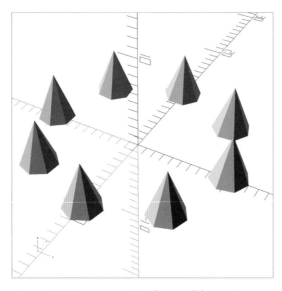

Figure 4-3: Ten cones, rotated around the z-axis, generated with a for loop

Inside the curly brackets, the loop creates a cone ❸, translates it 10 units along the x-axis ❷, and then rotates it by angle degrees ❶. The first cone is drawn when the value of the angle variable is 0, so it is not rotated at all. The value of the angle variable increases by 45 each time the loop is repeated, rotating each cone accordingly. The last value of the angle variable is 315, so the last cone drawn by the loop is rotated by 315 degrees around the z-axis.

Debugging for Loops with echo

Sometimes it's useful to examine the value of a variable as it changes during the repetition of a for loop. For instance, if you want to double-check your mental math, it can be easier to see the exact values being generated by the for loop. Use the echo function to print each successive value of a variable to the console window, and check the console window (Figure 4-4) to gather feedback about the execution of your OpenSCAD code:

```
for (x_position = [10:10:100]) {
    translate([x_position, 30, 0]) cylinder(h=4, r1=4, r2=4);
    echo("x:", x_position); //a good way to check your mental math
}
```

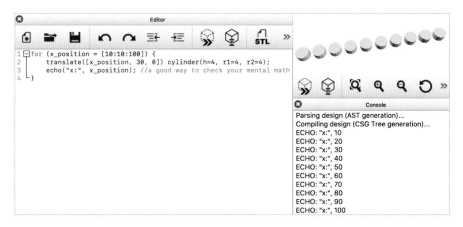

Figure 4-4: Console output generated with echo

The echo function is helpful for debugging your programs. For example, you can use it to visualize all the values of a variable that controls the number of times a for loop repeats. The echo function provides a useful way to gather feedback about your for loops, because it will print out every value generated by the for loop. Adding string labels (like "x:") to your console statements can help organize the console window output. Labels and variables in echo functions should be separated with commas (,).

Using Variables and Arithmetic

Variables are used in conjunction with for loops to keep track of a pattern created by the looping. You can either use the generated values directly, or you can perform arithmetic on them to produce more sophisticated repetitions.

In this section, you'll learn variable naming best practices, mathematical operations to perform on variables, and applications of variables within loops.

Naming Variables

Neither the x_position variable from the preceding for loop example nor the angle variable from Figure 4-3 is built into OpenSCAD. Those names were chosen to describe how the values are used in the design. The x_position variable describes the x-position of the cylinder, while angle describes the angle of rotation of the cone.

OpenSCAD allows you to name your variables however you want, as long as you don't include spaces or use any symbols other than letters, underscores, or numbers. Be sure to select a name that helps you remember a variable's purpose. This allows you to keep track of multiple variables in a design more easily, which can help tremendously when debugging errors or sharing your design.

Applying Mathematical Operations on Variables

To start exploring how OpenSCAD applies mathematical operations on variables, say you assign the values 10 and 3 to the following variables:

```
value1 = 10;
value2 = 3;
```

To perform mathematical operations like finding the sum, difference, product, quotient, or remainder of these values, OpenSCAD lets you use standard symbols.

OpenSCAD also respects the conventional order of operations that you are probably familiar with from math class. Assigning the result of each of these arithmetic operations to a variable will help you separate your calculation statements from your output statements:

```
sum = value1 + value2;
difference = value1 - value2;
product = value1 * value2;
quotient = value1 / value2;
remainder = value1 % value2;
```

Now, use the echo function to display the result of each mathematical operation (Figure 4-5). Each echo function uses a label to help identify which number is which in the console window.

```
echo("Addition:", sum);
echo("Subtraction:", difference);
echo("Multiplication:", product);
echo("Division:", quotient);
echo("Modulo:", remainder);
```

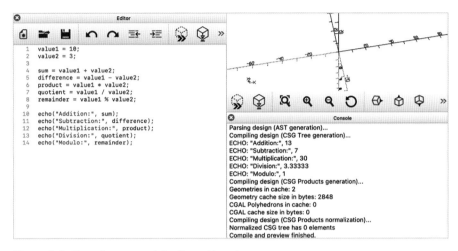

Figure 4-5: Console output of the five arithmetic operators

Using Math and Variables Inside for Loops

You can use arithmetic inside a for loop to make a single variable represent two patterns. The following design creates 13 spheres that are all generated by the same for loop (Figure 4-6):

```
for (faces=[3:❶1:15]) {
    ❷ $fn = faces;
    x_position = faces*10;
    translate([❸x_position, 0, 0]) sphere(r=5);
    ❹ echo("faces:", faces, "x-position:", x_position);
}
```

Figure 4-6: A succession of increasingly smoother spheres

Notice how the faces variable created by the for loop specifies both the number of faces used to render the sphere ❷ and the position of the sphere along the x-axis ❸. With each repetition of the for loop, the value of faces increases by one ❶, while the value of x_position is updated by multiplying the new value of the faces variable by 10. The echo function ❹ displays the changing values of faces and x_position. Figure 4-7 shows the console output.

```
ECHO: "faces:", 3, "x-position:", 30
ECHO: "faces:", 4, "x-position:", 40
ECHO: "faces:", 5, "x-position:", 50
ECHO: "faces:", 6, "x-position:", 60
ECHO: "faces:", 7, "x-position:", 70
ECHO: "faces:", 8, "x-position:", 80
ECHO: "faces:", 9, "x-position:", 90
ECHO: "faces:", 10, "x-position:", 100
ECHO: "faces:", 11, "x-position:", 110
ECHO: "faces:", 12, "x-position:", 120
ECHO: "faces:", 13, "x-position:", 130
ECHO: "faces:", 14, "x-position:", 140
ECHO: "faces:", 15, "x-position:", 150
```

Figure 4-7: The console output of a succession of increasingly smoother spheres

Using Arithmetic to Create Unique Patterns

In addition to using arithmetic to leverage the power of a for loop to progressively change characteristics of a shape, you can also use arithmetic to create interesting patterns. The following code generates a sequence of cylinders of increasing heights by using a quadratic pattern to increase the height of each cylinder (Figure 4-8):

```
for (❶x=❷[1:1:10]) {
    height = ❸x*x;
    x_position = ❹5*x;
    translate([x_position, 0, 0]) cylinder(h=height, r1=2, r2=2);
}
```

The preceding design uses a for loop to increase one variable, called x ❶, from 1 to 10 ❷. The x variable increases by one each time the loop repeats, so this loop repeats 10 times. This variable controls both the position along the x-axis and height of a series of cylinders. By creatively using arithmetic, you change the x position of the cylinder by 5 ❹ every time the loop repeats. The height of the cylinder grows at a different rate, by squaring the value of x every time the loop repeats ❸; this is known as *quadratic growth*.

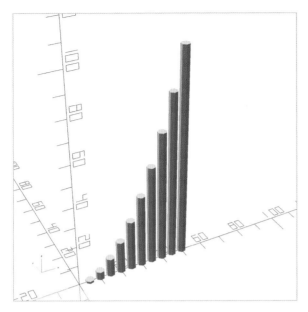

Figure 4-8: A succession of cylinders following a quadratic pattern to increase height

Using Nested Loops to Draw 2D and 3D Grids

OpenSCAD even lets you repeat a loop, so you can put a for loop inside another for loop. Whereas you can use one for loop to create a line of shapes, you can use a for loop inside another for loop to repeat that line of shapes to create a grid of shapes with only a few lines of code. This is called *nesting* the loops. The following design uses nested for loops to draw a grid of cylinders (Figure 4-9):

```
❶ for (y_pos = [10:10:50]) {
❷ for (x_pos = [10:10:100]) {
      translate([x_pos, y_pos, 0]) cylinder(h=4, r1=4, r2=4);
   ❸ echo("x:", x_pos, "y:", y_pos);
   } // x_pos loop
} // y_pos loop
```

The preceding code uses one loop to draw a line of 10 cylinders ❷. That for loop is repeated by the first for loop ❶, so the line of cylinders repeats. Two variables—that is, the x_pos and y_pos variables—work together to change both the x position and y position of the repeated cylinder. The inner loop repeats 10 times, while the outer loop repeats 5 times. This generates a total of 50 cylinders. The echo function is used to keep track of the changing values of both variables in the console window ❸. Notice that comments are used to indicate which bracket belongs to which loop. Commenting brackets isn't necessary but can be helpful when you have many curly brackets next to each other.

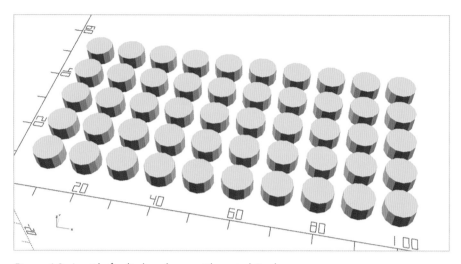

Figure 4-9: A grid of cylinders drawn with nested for loops

You now know how to generate 50 cylinders with four lines of code, which certainly beats writing a long list of 50 statements to generate each cylinder individually. This would be the perfect technique for drawing the many windows in a skyscraper.

Generating the Windows in a Skyscraper with Nested Loops

Listing 4-1 draws a building with 60 windows (Figure 4-10) by using nested for loops:

```
num_rows = 10;
num_cols = 6;

building_width = num_cols*5;
building_height = num_rows*6;

❶ difference() {
  ❷ cube([building_width, 10, building_height]);

  ❸ for (z = [1:1:num_rows]) {
    for (x = [0:1:num_cols-1]) {
    ❹ x_pos = x*5+1;
      z_pos = z*5;
      translate([x_pos, -1, z_pos]) cube([3, 3, 4]);
    } // x loop
    } // z loop
} // difference
```

Listing 4-1: Drawing a skyscraper with 60 windows by using nested for loops

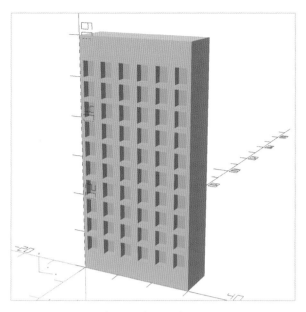

Figure 4-10: A window grid on a skyscraper

Listing 4-1 uses variables (named num_rows and num_cols) to control not only the number of windows, but also the width and height of the building. First, it draws a large cuboid to represent the building ❷. Next, it uses nested for loops to draw a grid of 60 cuboids ❸. Finally, the difference operation subtracts the cuboids from the larger building to create recessed windows ❶. Two variables (x_pos and z_pos) are used to calculate the specific x position and z position of each window prior to drawing the cuboid ❹.

Our organization of the code in Listing 4-1 makes it easy to change the skyscraper's characteristics. The variables num_rows and num_cols not only control the number of times the two loops repeat, but also set the width and height of the building, because the values of the building_width and building_height variables are dependent on the values of num_rows and num_cols. Making one change to either num_rows or num_cols will completely change the skyscraper's look. You'll learn more about the advantages of this sort of organization in the next chapter.

Triple Nesting to Create a 3D Grid of Shapes

You can also draw a 3D grid of shapes by adding another layer of nesting—that is, by putting a loop inside a loop, inside a loop—although this might take a while to render since it will generate a large number of shapes (Figure 4-11):

```
for (r = [0:15:255]) {
  for (g = [0:15:255]) {
    for (b = [0:15:255]) {
```

```
        translate([r, g, b]) color([r/255, g/255, b/255]) cube(5);
      } // b loop
    } // g loop
} // r loop
```

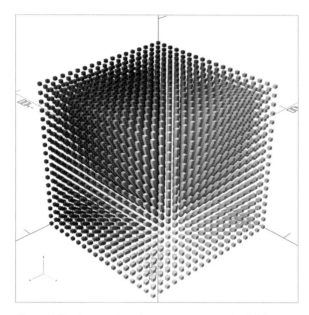

Figure 4-11: A nested for loop representing the RGB color space

This *triple nesting* essentially uses a third loop to repeat a grid of shapes. The preceding design uses three nested loops to draw a cube representing the RGB (red, green, blue) color space. The color transformation takes a 3D vector indicating the percentage of red, green, and blue light that should be represented in the color of the shape. Since RGB uses 255 as the maximum value, dividing by 255 results in a decimal between 0 and 1. The color transformation can be useful for debugging and organizing your designs, but it is not very useful for 3D printing, since the color of a 3D print depends entirely on the type of filament used. Thus, the color transformation is effective only in Preview mode and will not display in Render mode.

Summary

This chapter introduced the concept of looping, which lets you repeat statements without rewriting the same code. Looping lets you tell the computer to do all of the work of rewriting a statement over and over again. Variables are an important part of looping in OpenSCAD, although they are not exclusive to looping. Variables can also help you keep track of important values. Through the use of arithmetic operators, variables can act as important starting points for other variables, which is useful when you want to make changes to your design.

The following are some important tips for using loops:

- If you find yourself copying, pasting, and making minor changes to a repeating statement, consider generating the repetition with a loop.
- Use arithmetic to create sophisticated repetitions based on the pattern created by a loop.
- Give variables names that describe their purpose.
- Organizing all of your variables at the top of your program makes it easy to change your design.
- Use the echo function to output the value of a variable as a loop repeats. This can help you keep track of variables that are the result of complicated arithmetic.
- Label all echo function output so you can output several variables when you have nested loops.
- If you want to use variable values generated by a for loop in a text shape, remember to convert the number to a string with str (as mentioned in Chapter 3).
- The color transformation is useful for debugging in Preview mode, but it does not translate to Render mode or 3D printing.
- Comments are notes programmers leave to help explain their coding choices.
- OpenSCAD ignores comments, but humans use comments to help figure out what coding statements are trying to accomplish.

Before reading further, practice the skills you learned in this chapter by building each of the complex designs in Figure 4-12.

1. Regular prisms

2. Pyramid

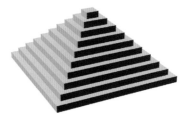

3. Magic coins

4. Flower

5. Grooved-edge coin

6. Stairs

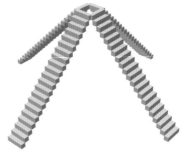

Figure 4-12: Practice drawing these complex designs.

Practice the skills you learned in Chapters 1 to 4 with the following three projects.

DETAIL TEST

Use loops to generate this detail test (Figure 4-13) for your 3D printer. This design helps you understand the lower limits of the printability of the smallest details on your designs. OpenSCAD will always render a virtual model with all of the details you've specified, but just because you can render fine details virtually doesn't mean every detail will be visible when you 3D-print your design. Some details are just too small to print.

Figure 4-13: Detail test

3D-Printing Tip for the Detail Test

In Chapter 1, you played around with varying the resolution of your 3D print by selecting High, Medium, or Low quality default settings. Print this detail test a few times, each time using your 3D printing preparation software to manually select a different layer height. Try to determine the smallest (close to 0.1 mm) and largest (close to 0.34 mm) layer heights for your printer; then make two prints of this design to see the effect of each layer height on the fine detail resolution of your prints.

TOWERS OF HANOI PUZZLE

Create a Towers of Hanoi puzzle using two loops (Figure 4-14). One loop should create the pegs, and one loop should create the discs. Remember to create holes in each disc that are slightly larger than the peg.

(continued)

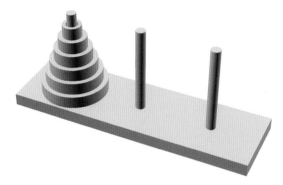

Figure 4-14: Towers of Hanoi puzzle

3D-Printing Tip for the Towers of Hanoi Puzzle

This design uses two loops: one for the discs and one for the bars. The picture represents how you might play the Towers of Hanoi game (which requires you to move all discs to the opposite bar by only stacking smaller discs on top of larger discs). The game begins in this configuration, with each disc stacked on a larger disc beneath it.

However, if you try to print your Towers of Hanoi game in this configuration, you won't be able to play! The discs would print as one solid unit. Once you have created your design, try to modify the loop that creates the discs so that they appear horizontally behind the game board. This will allow you to print the individual discs as separate units.

TIC-TAC-TOE GAME

Use loops to create a tic-tac-toe game (Figure 4-15).

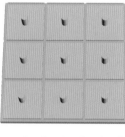

Figure 4-15: Tic-tac-toe game

3D-Printing Tip for the Tic-Tac-Toe Game

Don't forget that holes should have slightly larger diameters than pins; otherwise, the pieces of this tic-tac-toe game won't fit together. Also, notice how the game pieces are arranged relative to the game board. This arrangement will print just fine. However, you might want to use three different colors to print the Xs, Os, and game board. In that case, you can successively comment out sections of your code to render and download three different *.stl* files for printing with different filaments.

5

MODULES

In this chapter, you'll learn how to turn complex designs into more manageable components called modules. *Modules* are separate sections of code that organize a collection of stand-alone statements, and they're particularly useful for two reasons. If your code is long and complicated, using modules can break your code into smaller subsections, which helps make your code more readable. And if your design has duplicate or similar shapes, you can use a single module to define the shape, reducing the amount of code you need to write to create complex designs.

This chapter also describes how to use variables and parameters to customize your modules. Finally, we'll explain how to group similar modules into a separate file (often called a *library*) to make it easier to organize designs, share designs, and use designs others have created.

Simplifying Code with Modules

To understand how using modules might simplify your code, let's take another look at the code for drawing the cross-shaped cookie cutter (Figure 5-1) you built in Chapter 3.

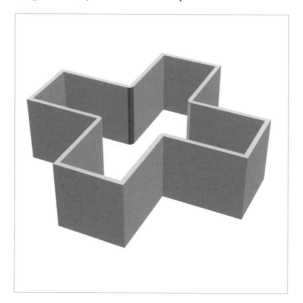

Figure 5-1: The cross-shaped cookie cutter

We've reproduced the code in Listing 5-1. Do you see any repeated code?

```
linear_extrude(30) {
    difference() {
        union() {
            square([100, 30], center=true);
            square([30, 100], center=true);
        }
        offset(-2) {
            square([100, 30], center=true);
            square([30, 100], center=true);
        }
    }
}
```

Listing 5-1: The original cross-shaped cookie cutter program

The cookie cutter is made by taking the difference of two crosses, so the square commands to create the cross shape are repeated twice. Duplicate code almost always causes problems, because any change you make to a shape's dimensions must be made twice (or however many times the code is duplicated). If you forget to change every instance, you'll need to spend time fixing it later, or worse, end up with lasting mistakes in your design.

To improve this design, you can use a module to create a cross shape, and then use that module to create each of the two crosses. The following pseudocode shows the syntax of a module definition:

```
module ModuleName() {
    // code used to define the new shape
}
```

Use the module keyword to start defining a new module. Then give the module a name that describes the new shape you are creating. Module names have the same restrictions as variable names, meaning you can only use lowercase and uppercase letters, underscores, or the digits 0 to 9. A good name should help readers understand what the module does without making them read the actual code that defines the module. Following the *ModuleName*, add an empty pair of parentheses followed by the code enclosed in curly brackets. The code you write inside the curly brackets is no different from any other OpenSCAD code.

The module definition will stand alone as a separate section of your design. So, defining a module won't actually draw the new shape. It's simply a recipe that describes how to create a shape. To see the shape, you must create it by inserting the module name into your design, just as you would to create any other shape. Here's the syntax for using a module:

```
ModuleName();
```

A module is an example of a programmer-defined shape. In fact, all the OpenSCAD commands you have used so far, including sphere, cylinder, and linear_extrude, are actually modules that are built into the language. An implied union operation occurs when shapes are combined within a module, so you can transform and combine the shape(s) generated by a module with any operation you've seen so far.

Write some new code for your cookie cutter by creating a cross module, as shown in Listing 5-2.

```
module cross()❶ {
    square([100, 30], center=true);
    square([30, 100], center=true);❷
}

linear_extrude(30) {
    difference() {
        ❸ cross();
        ❹ offset(-2) cross();
    }
}
```

Listing 5-2: The new cross-shaped cookie cutter program, improved with a module

Use the module keyword to start the definition of the new shape. Give it the name cross ❶ to describe the shape you're creating. In curly brackets following the name, enter the code for the shapes that define the cross ❷. Finally, tell OpenSCAD to draw the cross by using the module name followed by a set of parentheses ❸ ❹. Notice that you use the cross module twice, so you can subtract one cross shape from the other with the difference operation.

TIPS

Curly brackets are optional if they enclose a single shape. So this

```
offset(-2) {
    cross();
}
```

is the same as this:

```
offset(-2) cross();
```

And, a union of a single shape is the same as the shape itself, which means that this

```
union() {
    cross();
}
```

is the same as this:

```
cross();
```

Splitting Your Design into Multiple Files

Sometimes when creating a new design, you'll want to reuse a component from a previous project. A good way to organize this is to make the component into a module. Putting this module definition into a separate file will allow you to easily use it in both designs. Saving modules separately helps you find and reuse your new shapes in as many projects as you like, as well as easily share them with others. Also, if you make improvements to a module defined in a file that is used by several designs, those improvements will be applied the next time you open each design. Organizing module definitions into separate files is often called creating a *library*, especially when a new file has multiple related modules defined within it.

To learn how to save your module in a separate file, let's split the cross-shaped cookie-cutter design into two files. We'll use one file to define a cross shape, and then use that module in the second file to create a cookie cutter. First, create two empty OpenSCAD files: *cross-module.scad* and *cookie-cutter.scad*. Make sure you save the two files in the same folder so OpenSCAD can find the two files. Also, note that these filenames were chosen to clearly indicate the purpose of each file. Carefully choosing your filenames will help you organize your projects in the future, especially as you build more and more OpenSCAD projects.

In *cross-module.scad*, copy the module definition from Listing 5-2, including the curly brackets, and then paste it into the file you just created. Be sure to save *cross-module.scad* after you've pasted the code so that OpenSCAD can use the newest version when you connect the files. The new *cross-module.scad* file should contain only the following code:

```
module cross() {
    square([100, 30], center=true);
    square([30, 100], center=true);
}
```

Now in *cookie-cutter.scad*, remove the module definition and add the following line at the top of your file:

```
use <cross-module.scad>

linear_extrude(30) {
    difference() {
        cross();
        offset(-2) cross();
    }
}
```

Instead of typing the module definition in *cookie-cutter.scad*, the first line tells OpenSCAD to use code from *cross-module.scad*. This is what provides the definition for the cross shape.

The use keyword tells OpenSCAD to load the modules from a different file. The syntax for the use keyword is as follows:

```
use <path/to/filename.scad>
```

After the use keyword, add *angle brackets* (< >), and inside the angle brackets, specify the name of the *.scad* file you want to use. If the file you want to use is not in the same folder as your main design file, specify either the absolute or relative path to the file. A use statement allows you to use the module definitions from the file, but it will not immediately result in any shape being drawn.

Generating a preview of *cookie-cutter.scad* will now produce the same shape as in Figure 5-1. However, generating a preview of *cross-module.scad* will not produce any shape. That is because *cross-module.scad* currently only contains a definition of the cross module. In order to see what the cross shape looks like by generating a preview of *cross-module.scad*, you need to add a statement to draw the cross:

```
cross();

module cross() {
    square([100, 30], center=true);
```

```
    square([30, 100], center=true);
}
```

Adding Parameters to Your Modules

Because shapes come in different sizes, you'll likely want your modules to allow for some variation. You already know that built-in OpenSCAD modules, like sphere, can take a parameter, such as sphere(r=30);, where the parameter specifies the sphere's radius. You can add such parameters to your own modules as well.

The following pseudocode shows the full syntax for specifying a module, including parameters:

```
module ModuleName(parameterName = defaultValue, ...) {
  // statements used to define the shape
}
```

Instead of leaving the parentheses after the module definition empty, add a *parameterName*, which is a placeholder for a value that you'll provide whenever you use the module. You can also give each parameter a *defaultValue*, so if the user of a module doesn't specify a value for a parameter, the module will use the default value instead. Providing a default value allows people to use the module without having to specify all parameters, which can be beneficial when experimenting with a module, or it can hide distracting details when the default value is a common choice. To create multiple parameters, specify multiple parameter names, separated by commas, and be sure to give each parameter a different name.

You may have noticed that parameters look a lot like variables. In fact, inside a module, parameters behave as if they were variables. It's good practice to give parameters names that describe their purpose. As with variables and module names, parameter names can only include letters, underscores, or numbers.

Listing 5-3 shows how to add parameters to the cross module:

```
module cross(width=30, length=100) {
    square([length, width], center=true);
    square([width, length], center=true);
}
```

Listing 5-3: Defining the cross module with parameters

Inside the parentheses, you add the width and length parameters, which define the width and length of each arm of the cross.

To create a cross shape with the cross module, provide specific values for each parameter each time you use the module, as shown in Listing 5-4.

```
use <cross-module.scad>

linear_extrude(30) {
    difference() {
```

```
        cross(20, 100);
        offset(-2) cross(20, 100);
    }
}
```

Listing 5-4: Specifying values for the cross module

The order of the numbers indicates which should be interpreted as the width of the cross and which should be interpreted as the length of the cross. Since the `width` parameter comes first in the definition of the module, the first number in the parentheses is assigned to the `width` parameter, and the second number is assigned to the `length` parameter.

OpenSCAD also allows you to name your parameters explicitly when you use a module, which can be helpful when you create a shape with a large number of parameters (and keeping track of the order becomes unwieldy):

```
cross(width=20, length=100);
```

When you use a module and name your parameters, the order of the parameters is not important. Switching the order of the length and width parameters does not affect the appearance of the shape:

```
cross(length=100, width=20);
```

Now the module is truly dynamic; you can use it to create cookie cutters of any size (Figure 5-2).

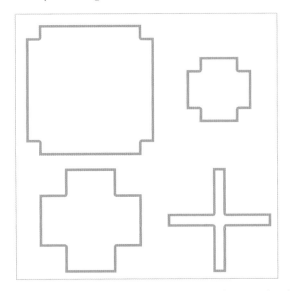

Figure 5-2: A variety of cookie cutters, each created with different parameters

Building a LEGO Brick

In this section, we'll walk through a complex modeling project that uses parameters, modules, and for loops in a single design. You'll design a LEGO brick shape that has two studs in one direction and any number of studs in the other direction. *Studs* are the small bumps on the top of a LEGO brick that fit into other LEGO bricks to hold them together. Figure 5-3 shows a LEGO brick with two rows and four studs per row.

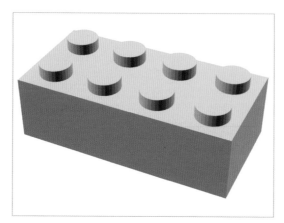

Figure 5-3: A LEGO brick with a 2×4 grid of studs

Before coding a complicated design like this, sketching a few hand-drawn versions of your shape can help you gain a firm understanding of the dimensions and patterns that exist within the shape (Figure 5-4).

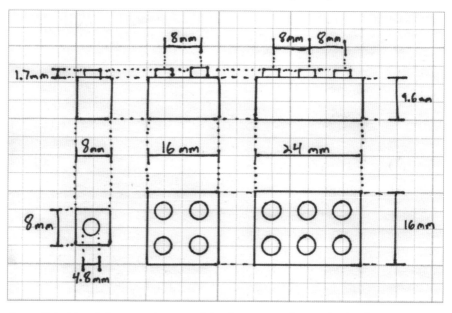

Figure 5-4: A hand-drawn exploration of the dimensions of LEGO bricks of various sizes

The dimensions of LEGO bricks are readily available online. We've taken our dimensions from Wikipedia:

- The height of a brick is 9.6 mm.
- The height of a stud is 1.7 mm.
- The diameter of a stud is 4.8 mm.

Each new stud adds 8 mm to the width of the brick to accommodate not only the diameter of a stud, but also the empty space surrounding a stud. The length of a brick is also dependent on the number of studs. You'll generate only bricks with two rows of studs, which implies a fixed brick length of 16 mm for this example.

Exploring a variety of hand-drawn LEGO shapes makes it easier to identify the OpenSCAD statements necessary for defining a LEGO brick module.

Listing 5-5 defines a LEGO brick module.

```
module LEGObrick(studs_per_row=4) {
    $fn=30;

    width = studs_per_row * 8;

    cube([width, 16, 9.6]);

    for (x_position=[4 : 8 : width-4]) {
        translate([x_position, 4, 1.7]) cylinder(h=9.6, d=4.8);
        translate([x_position, 12, 1.7]) cylinder(h=9.6, d=4.8);
    }
}

LEGObrick(4);
```

Listing 5-5: Drawing a LEGO brick with modules

Start by creating a module named LEGObrick with a studs_per_row parameter. This parameter represents the number of studs along the top of the LEGO brick, which determines the overall width along the x-axis of the brick. LEGO bricks come in different sizes, so this parameter will be useful as a way to reuse the same module to draw a variety of brick sizes. We chose to set a default value of 4 studs per row, but this is an arbitrary choice.

A variable called width is created to keep track of the overall width of the brick, which is based on studs_per_row. Each additional stud increases the width of the brick by 8 mm:

```
width = studs_per_row * 8;
```

Other dimensions of the LEGO brick remain fixed, unrelated to the number of studs per row:

```
cube([width, 16, 9.6]);
```

A for loop is used to draw each repeated stud in its proper position:

```
for (x_position=[4❶ : 8❷ : width-4❸]) {
    translate([x_position, 4, 1.7]) cylinder(h=9.6, d=4.8);
    translate([x_position, 12, 1.7]) cylinder(h=9.6, d=4.8);
}
```

Inside the for loop, the variable x_position keeps track of the x position of each stud. The first stud is centered at x = 4 mm ❶, and each additional stud is positioned 8 mm ❷ away from the previous stud. Similarly, the last stud in each row is centered 4 mm from the overall width of the brick ❸. Two rows of studs are drawn with the exact same values on the x-axis. Since we're restricting ourselves to just two studs on the y-axis, it's easier to position the rows explicitly at y = 4 mm and y = 12 mm instead of using a second loop.

The LEGObrick module is now complete, which means you can use it to create LEGO bricks of various sizes, like the ones in Figure 5-5.

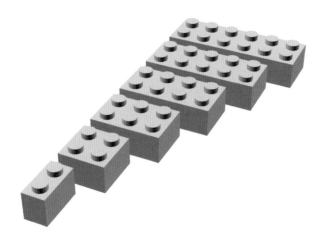

Figure 5-5: A variety of LEGO bricks created with the same LEGObrick module

This module is only a simplified design of a LEGO brick, however; it won't function as a real brick would, because the current design doesn't include an interior mechanism on the bottom of the brick for snapping bricks together. We leave that as a challenge for you.

You might have noticed a new use of the cylinder module to generate the LEGOBrick module:

```
cylinder(h=9.6, d=4.8);
```

This version of cylinder uses a single d parameter to indicate that both faces of the cylinder should have the same diameter. In past chapters, we would have drawn a cylinder with two separate parameters to indicate the radius of individual faces:

```
cylinder(h=9.6, r1=2.4, r2=2.4);
```

OpenSCAD provides four alternative ways of using a cylinder module. Each method uses a different combination of named parameters:

```
cylinder(h, r|d, center)
cylinder(h, r1|d1, r2|d2, center)
```

The | indicates that you can either use r or d, depending on whether you prefer using the radius or diameter to define cylinder faces. If both faces have the same size, it can be easier and less error-prone to define the size of both faces just once. Otherwise, you can use two different parameters to indicate the size of each face. Which version you use is up to you!

In prior chapters, we used a single version of the cylinder module to reduce the number of new OpenSCAD commands needed to draw both cylinder and cone shapes. Alternative versions of the cylinder module are introduced in this chapter to illustrate the variety of choices you have in regard to choosing and naming parameters. Alternative versions of cylinder and other OpenSCAD modules are listed in Appendix A: OpenSCAD Language Reference.

Sharing and Collaborating

If you save your modules in separate files, you can reuse your new shapes in multiple designs, as you saw earlier in this chapter. Keeping your modules separate also allows you to share common design components with other people or use other people's components instead of building everything yourself. Splitting a design into multiple modules allows you to collaborate more easily.

Let's walk through a possible collaboration. Say you and a friend want to work together to make a 3D animation of a LEGO castle. To save time, you decide to split the design into two tasks that can be completed in parallel using two different computers. Your friend decides to design a module that will draw a LEGO brick shape, while you will be in charge of designing a castle that is made from LEGO brick shapes.

You and your friend first decide what the LEGO module should look like. You agree on a name for the module (LEGObrick), any necessary parameters and their defaults (studs_per_row), with a default of three studs, and the basic shape and size of each brick (24 × 16 × 9.6 mm for a 3×2 brick). Your friend then goes off and builds a simple version of the LEGObrick module in a file called *LEGObrick-module.scad*, shown in Figure 5-6:

```
LEGObrick();
module LEGObrick(studs_per_row=3) {
    cube([24, 16, 9.6]);
}
```

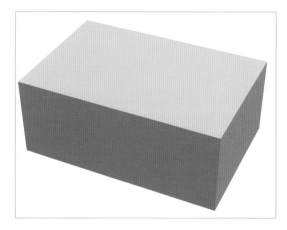

Figure 5-6: A simple version of the LEGObrick module

Even though the LEGObrick module isn't complete (this simple version of the module doesn't have studs yet), you can still use it as a building block to start creating the castle design in a file called *castle-wall.scad*, as shown in Figure 5-7.

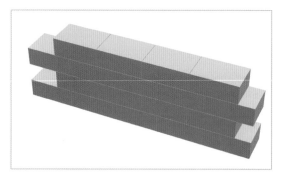

Figure 5-7: A wall of the LEGO castle that uses a basic version of the LEGObrick module

Meanwhile, your friend keeps working on the LEGObrick module, and every time it improves, your friend shares their new version of *LEGObrick-module.scad* with you. Because OpenSCAD designs are plaintext files (with a *.scad* extension), you can share them by emailing the files as attachments, copying and pasting OpenSCAD code directly from email or other documents, or by using more advanced services like GitHub to make designs public. 3D design-sharing websites also exist. One of the more popular ones, which supports OpenSCAD directly, is Thingiverse (*https://thingiverse.com/*).

Every time your friend shares an updated version of *LEGObrick-module .scad*, you replace your old version of the file with the new version. Your castle design in *castle-wall.scad* will update to use the newest definition of LEGObrick each time you Preview or Render your castle code. Over time, your design may look more like the one shown in Figure 5-8.

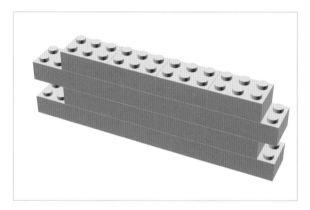

Figure 5-8: Building the castle by using the updated
LEGObrick module

This collaboration strategy saves you time, because you and your friend can work on the LEGObrick module and castle design simultaneously. You don't have to wait for your friend to finish a part before you can make progress on your own part, while your friend can see how small changes in their module design effect the overall castle design.

Summary

In this chapter, you learned how to organize your design into smaller logical components by using modules, which can make your OpenSCAD code more readable, facilitate collaboration, and help you customize your design.

When using modules, remember these key concepts:

- Module definitions have three parts: a name, a parameter list, and a body.
- The body of the module contains a collection of OpenSCAD statements that define the unique shape of the module.

- To use a module, create the shape by using the name of the module in your design. If your module isn't showing up, check that you're actually using the name of the module in your code statements; it's possible you've only defined the module.

- When designing a module, choose module names and parameters that obviously describe their purpose, so someone using your module won't need to read your module definition to know what it does. This can also help you later if you have forgotten the module's details.

- Parameters are useful for specifying a module's characteristics. Identifying which variables should be included as parameters is an important part of designing a module.

- Specifying default values for parameters is a useful way to make some parameters optional.

- Separating your module definition into other files helps you use the module in other OpenSCAD designs. You can also group related modules into a library. As with modules and variables, choose filenames that adequately describe their purpose.

- Connecting your design to a module with the use keyword won't immediately add new shapes to your design. You have to explicitly use the module in your code to see the new shape.

- It's common practice to draw the shape defined by a module at the top of a module definition file. This is helpful for testing purposes.

Try searching online for examples of OpenSCAD modules to see more examples of user-defined shapes. You can learn a lot by inspecting and tinkering with other people's solutions, especially when it comes to figuring out which parameters to include.

Practice using modules to define each of the shapes in Figure 5-9. We've suggested a few parameters for each module. Feel free to create modules with a different list of parameters.

1. Ring with hole_diameter, height, and thickness parameters

2. Pencil holder with diameter, height, and thickness parameters

3. Pencil cap with letter and pencil _diameter parameters

4. Name tag with name, length, width, and height parameters

5. Bag hook with hole_radius, height, and thickness parameters

6. Box with length, width, height, and thickness parameters

Figure 5-9: Practice using modules to create these designs.

Now try your hand at these two projects.

SKYSCRAPER

Design a module that generates a variety of skyscrapers (Figure 5-10). Control the design of your skyscraper by using parameters to determine its length, width, height, and number of windows. Don't forget to draw windows on the other side of the skyscraper!

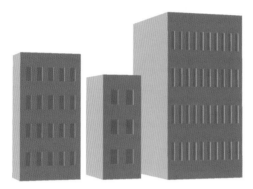

Figure 5-10: A variety of skyscrapers generated with the same module

LEGO LIBRARY

Build a library of creative brick modules by adding definitions for multiple brick types to the same file. Use parameters to control the length of each brick. Create corner bricks, wall bricks, and double-sided bricks (Figure 5-11). Try to create your own unique brick type to add to the library.

Figure 5-11: A variety of corner, wall, and double-sided LEGO bricks

6

DYNAMIC DESIGNS WITH IF STATEMENTS

In this chapter, you'll learn how to use `if` statements to create OpenSCAD designs that respond differently to a variety of conditions. This powerful programming tool gives your designs the power to choose from several options and execute only certain lines of code. As a result, you can create dynamic designs that adapt to changing circumstances. For instance, you can use `if` statements to reconfigure a design quickly for 3D printing.

As an example project, here you will learn to use `if` statements to vary the length of tick marks on a ruler to indicate inch, half-inch, and quarter-inch increments depending on the position of the tick mark. You'll also learn how to use random numbers to vary repeated shapes in order to create a more organic variety of design characteristics.

Using if Statements

An if statement uses a Boolean expression (an expression that evaluates to either true or false) to compare two values, then determines whether to execute code based on that comparison. If the Boolean expression in an if statement evaluates to true, the indicated code statements are executed. Otherwise, the statements are skipped entirely. The Boolean expression describes a condition that must be satisfied in order for the indicated statements to be added to the design.

The following shows if statement syntax:

```
if (<boolean expression>) {
  // code that is executed only when the boolean expression is true
}
```

Listing 6-1 is a variation on the skyscraper design created in Chapter 4. This new version uses if statements to decide where to place windows and doors in the skyscraper (Figure 6-1).

```
num_rows = 10;
num_col = 6;

building_width = num_col * 5;
building_height = num_rows * 6;

difference() {
  cube([building_width, 10, building_height]);

  for (❶ z = [1:1:num_rows]) {
    for  (x = [0:1:num_col-1]) {
    ❷ if (z == 1) {
      ❸ translate([x*5+1, -1, -1]) cube([3, 3, 8]); // door
      }
    ❹ if (z > 1) {
      ❺ translate([x*5+1, -1, z*5]) cube([3, 3, 4]);  // window
      }
    }
  }
}
```

Listing 6-1: Using if statements to insert doors and windows depending on floor number

Figure 6-1 shows a skyscraper with doors on the first floor and windows on every subsequent floor. Two for loops in Listing 6-1 create the rows and columns of windows and doors in this design. The z variable ❶ controls the vertical position of each row. Next, two if statements use those z values to decide whether to add a window or a door to the design. If z equals 1 ❷, a large door is added to the design ❸. If z is greater than 1 ❹, a small window is added to the design ❺.

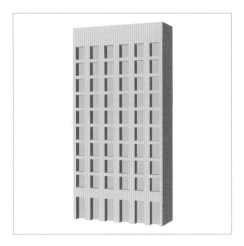

Figure 6-1: A grid of windows on a skyscraper, with a row of doors

We'll evolve this skyscraper design throughout the rest of the chapter. However, you might feel that a skyscraper is not a skyscraper without more details, especially on the other sides of the building. We totally agree and leave the exercise of adding more detail to this simple design as a challenge to the reader.

Defining Complex Conditions

You can use an if statement to evaluate many types of conditions by utilizing a combination of six Boolean operators and one of two logical operators. In addition, you can specify a default scenario (which is executed if the specified condition is false) by connecting an else statement with an if statement. Finally, you can connect several related conditions together by using an else if statement.

Choosing Boolean Operators

OpenSCAD uses six Boolean operators to evaluate the content of variables within a Boolean expression. Each of these operators will result in true if the comparison is valid, and false if the comparison is not valid:

- ‹   less than
- ›   greater than
- ‹=   less than or equal to
- ›=   greater than or equal to
- ==   equal to
- !=   not equal to

The symbols used for many of these Boolean operators are probably familiar to you from math class. OpenSCAD (as with most other programming languages) changes the symbols a bit so that you can easily type them on a keyboard. For instance, you're probably used to seeing the ≤ symbol to indicate less than or equal to. However, programming languages commonly use <= instead. In the same way, >= replaces ≥, and != replaces ≠. Finally, be sure not to confuse == with =. Because the single equal sign already has a use (assigning a value to a variable), Boolean expressions use the double equal sign (==) to test whether two values are "equal to" each other. For example, Listing 6-1 tests for the equality of two values by using the equals (==) operator.

This collection of Boolean operators provides many choices for evaluating variables to determine whether a condition is true or false. You can now write a loop that generates different shapes depending on the number of times the loop has repeated. As you will see later, you can also specify that you'd like to draw a shape only if a certain condition is not true. Using Boolean operators in an if statement allows you to create dynamic designs with a relatively small number of statements.

Using Logical Operators to Combine Boolean Expressions

Additionally, you can combine multiple Boolean expressions with one of two logical operators: && (which stands for *and*) and || (which means *or*).

If you use the && operator, all conditions need to be true in order for the indicated statements to execute. If you use the || operator, at least one of multiple conditions needs to be true. For a better sense of how the && operator works, consider the following:

```
if (x > 10 && y <= 20) {
  translate([x, y, 0]) cube([3, 4, 3]);
}
```

This code segment draws a translated cube only when x is greater than 10 *and* y is less than or equal to 20.

Now consider this if statement that uses the || operator:

```
if (x > 10 || y <= 20) {
  translate([x, y, 0]) cube([3, 4, 3]);
}
```

A translated cube is drawn when *either* x is greater than 10 *or* y is less than or equal to 20. Only one of the Boolean expressions connected by an *or* operator needs to evaluate to true in order for the shape to be drawn. The cube will also be drawn if both Boolean expressions connected by the *or* operator are true.

Following an Expanded Order of Operations

You can construct complex Boolean expressions that involve many arithmetic, Boolean, and logical operators. As in math class, where you learn to perform multiplication *before* addition, even if addition comes first in the arithmetic expression, OpenSCAD evaluates expressions following a well-defined order of operations:

1. ( )
2. ^
3. *, /, %
4. +, -
5. <, >, <=, >=
6. ==, !=
7. &&
8. ||

Operators at the same level in the order of operations are performed according to the order of their occurrence in the expression as it is read from left to right. Otherwise, operators at the top of this list have a higher precedence and are calculated prior to operators at the bottom of the list, even if that means the expression is calculated from the inside out.

Making Two-Way Choices with if...else Statements

A basic if statement executes a section of code only when the Boolean condition is true. To execute alternate code when the Boolean condition is false, attach an else statement to an if statement. An if...else statement creates a two-way branch in your code, allowing you to execute different collections of statements for each truth condition.

Consider the following if...else syntax:

```
if (<boolean expression>) {
  // code that is executed only when the boolean expression is true
}
else {
  // code that is executed only when the boolean expression is false
}
```

If the Boolean expression in the if statement is true, the first group of statements is executed. If the Boolean expression in the if statement is false, the statements contained within the else section is executed. The else section of an if statement is often called the *default* condition, because it describes what should happen when the condition specified in the if statement is false. An else statement is an optional extension to an if statement and is appropriate for *mutually exclusive* branching scenarios, where there is no possibility that you want to include both sections of code in your design.

You could easily redesign Listing 6-1 by using an else statement. The skyscraper in Figure 6-1 has exactly one row of doors. All of the remaining rows will have windows. Because the for loop should sometimes draw a door and all other times draw a window, you could rewrite the if statement like this:

```
num_rows = 10;
num_col = 6;
```

```
building_width = num_col * 5;
building_height = num_rows * 6;

difference() {
  cube([building_width, 10, building_height]);

  for (z = [1:1:num_rows]) {
    for (x = [0:1:num_col-1]) {
      if (z == 1❶) {
      ❷ translate([x*5+1, -1, -1]) cube([3, 3, 8]); // door
      }
      else {
      ❸ translate([x*5+1, -1, z*5]) cube([3, 3, 4]);  // window
      }
    }
  }
}
```

If the Boolean expression z == 1 ❶ is true, OpenSCAD draws a door ❷. If the Boolean expression is false, OpenSCAD draws a window ❸.

Using Extended if Statements

An *extended if statement* attaches a condition to an else statement to create an ordered collection of related decisions. OpenSCAD evaluates the Boolean expressions in an extended if statement in order until one of the expressions evaluates to true. You can optionally include an else statement at the end of an extended if to provide a catchall default option in case all of the decisions evaluate to false.

The syntax for an extended if statement looks like this:

```
if (<boolean expression>) {
  // code that is executed only when the boolean expression is true
}
else if (<boolean expression>) {
  // code that is executed only when the first boolean expression is false
  // and the second boolean expression is true
}
else {
  // optional default scenario
  // code that is executed only when both boolean expressions are false
}
```

You can add as many else if statements as needed to describe any number of mutually exclusive design possibilities, which is particularly useful when you want to ensure that exactly one of many related outcomes should happen. Each Boolean expression in the extended if statement is evaluated in order until one is found that evaluates to true. Only the code section for that Boolean expression is executed, while the remaining sections are skipped. If no Boolean expressions are true, the code specified in the optional else section (if provided) is executed. Because the else section describes the default possibility, it must be included at the end of an extended if statement.

Listing 6-2 uses an extended if statement to draw tick marks of various sizes on a ruler (Figure 6-2). This design creates an inch-based ruler with tick marks at three repeating intervals: one inch, a half inch, and a quarter inch. Tick marks at inch intervals are the longest, followed by slightly shorter half-inch tick marks, and even shorter quarter-inch tick marks. OpenSCAD is unit-less, so this design uses basic proportionality to divide every inch on the ruler into four equal-sized "gaps." It is intended to be resized to its exact width in your 3D-printing preparation software just prior to printing.

```
ruler(5);

module ruler(inches) {
❶ gap_size = 1; // 1 unit per quarter inch
  total_marks = 4 * inches; // inch, half inch, quarter inch

  width = gap_size * total_marks;
  length = 4 * gap_size;
  height = 0.5 * gap_size;

  mark_width = 0.25 * gap_size;
  mark_height = 1.5 * height;

  // main ruler body
  difference() {
    cube([width, length, height]);
    translate([width-gap_size, length-gap_size, -0.5])
      cylinder(h=height+1, r=0.15*length, $fn=20);
  }

  // tick marks
❷ for(t = [1:1:total_marks-1]) {
    mark_x = gap_size * t - 0.5 * mark_width;

❸   if (t%4 == 0) { // inch marks and number labels
      translate([gap_size * t, 0.65 * length, 0])
        linear_extrude(mark_height)
          text(str(t/4), size=gap_size, halign="center");
      translate([mark_x, 0, 0])
        cube([mark_width, 0.5 * length, mark_height]);
    }
❹   else if (t%2 == 0) { // half-inch marks
      translate([mark_x, 0, 0])
        cube([0.75 * mark_width, 0.25 * length, mark_height]);
    }
❺   else { // quarter-inch marks
      translate([mark_x, 0, 0])
        cube([0.5 * mark_width, 0.125 * length, mark_height]);
    }
  }
}
```

Listing 6-2: Using extended if statements to differentiate tick mark sizes on a ruler

Figure 6-2: A five-inch ruler

First, a collection of variables is defined to help us organize our design ❶: gap_size indicates that one OpenSCAD unit will represent the width taken by a single quarter-inch gap between tick marks, and total_marks keeps track of the total number of tick marks needed (according to the inches parameter of the ruler module). We'll need four tick marks per inch as we'll include marks at the inch, half-inch, and quarter-inch intervals. The other variables relate the proportionality of various features of the ruler to these two initial choices. Organizing the module variables in this manner allows you to quickly update your design in the future. For instance, you might decide to make a longer ruler in your next version. This change could easily be accomplished by making a single change: the calculation that determines the value of length variable.

The for loop ❷ draws something for every needed tick mark, except for the first and last tick marks, which are meant to be inferred (as they are the beginning and end of the ruler). The t variable in the for loop keeps track of the number of tick marks being drawn, while mark_x is used to keep track of the location of each new tick mark along the x-axis. The first Boolean expression ❸ tests whether the t variable is divisible by 4 (remember, % calculates the remainder). If this condition is true, the longest tick mark is added to the design to indicate an inch interval. If the t variable isn't divisible by 4, the second Boolean expression ❹ tests whether it is divisible by 2. And if it is, the second-longest tick mark is added to the design to indicate a half-inch mark. Only if the t variable isn't divisible by either 4 or 2 will the shortest tick mark be added to the design ❺ by the else statement.

Notice the careful ordering of the decisions used in this extended if statement. The for loop produces a series of numbers that are each evaluated by the extended if statement: 1, 2, 3, 4, 5, 6, 7, 8, and so on. Numbers like 4, 8, and 12 are divisible by both 4 and 2, so which condition should be executed? Extended if statements evaluate each decision in order, executing only the code contained in the if statement with the first Boolean expression that is true. Even though some numbers are divisible by both 4 and 2, the second decision ❸ is evaluated only if the first expression ❷ is false. Thus, only one tick mark is drawn for each value of t in the for loop. This is an example of a mutually exclusive scenario: one, and only one, of the three tick mark lengths should be drawn for each value of t.

Using Nested if Statements

Placing an if statement inside another if statement is a way to guarantee that a Boolean expression should be considered only if another Boolean expression is true. At a basic level, a nested if statement can replace the && operator:

```
if (x < 8 && y == 10) {
  // code that is executed only when both boolean expressions are true
}
```

So you could rewrite the preceding code with a nested if statement:

```
if (x < 8) {
  if (y == 10) {
    // code that is executed only when both boolean expressions are true
  }
}
```

It's probably easiest to use the && operator for simple combinations of Boolean expressions that all need to be true for satisfying certain design conditions. However, using nested if statements can be easier when you want to test the outcome of multiple Boolean expressions that can either be true or false:

```
if (x < 8) {
  if (y == 10) {
    // code that is executed only when both x < 8 and y == 10
  }
  else if (y < 10) {
    // code that is executed only when both x < 8 and y < 10
  }
  else {
    // code that is executed only when both x < 8 and y > 10
  }
} else {
  if (y == 10) {
    // code that is executed only when both x >= 8 and y ==10
  }
  else {
    // code that is executed only when both x >= 8 and y !=10
  }
}
```

It's usually possible to describe complex conditions using a variety of combinations of Boolean operators, logical operators, extended if statements, and nested if statements. Often, the best choice is the combination of conditions that makes the most sense to the person creating the design.

Useful Applications of if Statements

You should include an if statement whenever you want your OpenSCAD design to vary according to a specific condition. The following situations are examples of when you may want to use if statements in your projects.

Setting Up a Design Mode and Print Mode

Consider the Towers of Hanoi project from Chapter 4. When designing the series of stacking discs, it was convenient to visualize the discs stacked vertically on one of the pegs. However, that configuration is not the best for 3D-printing the design, because the disks are resting on top of each other, and you wouldn't want all the disks to be printed as one piece.

A useful technique is to create two versions of your design: one configuration for visualizing the final result and one for 3D-printing it. Use *design mode* to build your design in a way that is easy to visualize, and use *print mode* to reorganize the same design in a way that is better for 3D printing.

Listing 6-3 incorporates these two design configurations; Figure 6-3 depicts print mode.

```
$fn = 100;
mode = "print"; // or "design"

cube([200, 60, 10], center=true);

for (x = [-60:60:60]) {
  translate([x, 0, 5]) cylinder(h=70, r=4);
}

❶ if (mode == "design") {
    for (d = [2:1:7]) {
      translate([-60, 0, 10 + (7-d)*10]) disc(d*4, 5);
    }
  }
❷ else if (mode == "print") {
    for (d = [2:1:7]) {
      if (d > 4) {
        translate([60*d - 350, 60, 0]) disc(d*4, 5);
      }
      else {
        translate([60*d - 200, 100, 0]) disc(d*4, 5);
      }
    }
  }

module disc(disc_radius, hole_radius) {
  difference() {
```

```
        cylinder(h=10, r=disc_radius, center=true);
        cylinder(h=11, r=hole_radius, center=true);
    }
}
```

Listing 6-3: Using if statements to differentiate print mode and design mode

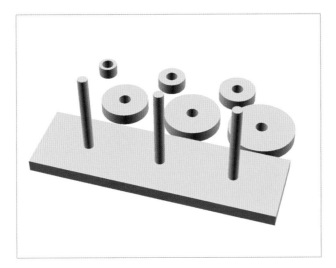

Figure 6-3: A Towers of Hanoi project set up for printing

Listing 6-3 uses a variable named mode and an extended if statement to decide whether to draw the shapes in "print" mode or "design" mode. If mode == "design" ❶, the disks are displayed vertically, stacked on top of one another, which makes it easy to visualize and check for correctness. If mode == "print" ❷, the discs are arranged in two horizontal rows, which is an appropriate setup for 3D printing. This differentiation allows you to quickly switch back and forth between the two configurations. When you are ready to print, all you need to do is change the value of mode to "print" to make the design change automatically.

Using Random Numbers as a Design Element

Random numbers are a fun way to add unpredictable elements to your designs, which is convenient when a design has features that you want to be similar, but not exactly the same. For instance, you could easily use random numbers to generate an entire city of skyscrapers, all with different heights, widths, and numbers of windows.

When you roll a six-sided die, you can expect that one of the six values (1, 2, 3, 4, 5, 6) on the die will be the result of the roll, but you can't predict the exact outcome. A similar procedure happens with the rands function. You can be certain that a decimal value within a specified range will be picked without knowing exactly which value will be picked until the statement is executed.

Use the mathematical rands function to generate random numbers. The following line of code picks two random decimal numbers between 0 and 1:

```
number_list = rands(0, 1, 2);
```

The first parameter you pass to rands specifies the lowest decimal number that the random number generator can choose. In this case, the lowest possible number is 0.0. The second parameter specifies the highest possible number, which is 1.0 for this example. The third parameter, 2, specifies how many numbers will be picked. The variable number_list remembers the generated list of random numbers so you can use them later.

The following code segment chooses three random numbers from 10 to 20, then stores the list of three numbers in a variable called number_list. Each random number in the list is then printed to the console window with the number_list variable, followed by the position of each number in the list within square brackets ([ ]). As with most programming languages, OpenSCAD considers the first element in a list to be in position [0]:

```
number_list = rands(10, 20, 3);

echo(number_list[0]);
echo(number_list[1]);
echo(number_list[2]);
```

Every time you preview this code, you will see a different combination of three randomly chosen decimal numbers from 10 to 20 printed to the console.

The rands function can choose any decimal number within the range you provide, but sometimes it's convenient to restrict a design to working only with integers (that is, numbers without decimals). If your design needs to pick a random integer within a certain range, the mathematical round function can be used to map randomly generated decimals to integers. The round function examines the decimal extension of a number to decide whether the decimal number should be "rounded up" or "rounded down" according to whether the decimal extension is >= 0.5 or < 0.5, respectively:

```
number_list = rands(9.5, 20.49, 3);

echo(round(number_list[0]));
echo(round(number_list[1]));
echo(round(number_list[2]));
```

Every time you run this code, you will see a different combination of three integers from 10 to 20 printed to the console because of the use of the mathematical round function in each echo statement. Notice that the first two parameters of the rands function have been changed to 9.5 and 20.49 in order to ensure that each integer in the original range (that is, 10, 11, 12, 13, 14, 15, 16, 17, 18, 19, or 20) is picked an approximately equally likely number of times. Because we wouldn't want to allow for a random choice of 20.5 and have it rounded up to 21, we use 20.49 as the highest possible value that can be generated. This produces a slightly lower possibility of 20 being randomly generated as compared to the other integer values in the range, but the difference is very small.

Random numbers are a useful way to generate design elements only a certain percentage of the time. For instance, you could modify your skyscraper design from the preceding chapter so that 50 percent of the time, the skyscraper includes a water tower on top of the roof.

Listing 6-4 draws the same simple skyscraper from Listing 6-2. This new version of the design sometimes includes a water tower to the top of the building (Figure 6-4).

```
num_rows = 10;
num_col = 6;

building_width = num_col * 5;
building_height = num_rows * 6;

difference() {
  cube([building_width, 10, building_height]);

  for (z = [1:1:num_rows]) {
    for (x = [0:1:num_col-1]) {
      if (z == 1) {
        translate([x*5 + 1, -1, -1]) cube([3, 3, 8]);
      }
      else {
        translate([x*5 + 1, -1, z*5]) cube([3, 3, 4]);
      }
    }
  }
}

❶ draw_tower = rands(0, 1, 1);

❷ if (draw_tower[0] < 0.5) {
    translate([building_width/6, 5, building_height])
      watertower(building_width/4);
}

module watertower(width) {
  $fn = 20;
  cylinder(h=5, r=width/2);
  translate([0, 0, 5]) cylinder(h=5, r1=width/2, r2=0);
}
```

Listing 6-4: if statements and random numbers to sometimes draw a water tower

After drawing a basic building, the design generates a list with a single random number between 0 and 1 ❶. This list is stored in the draw_tower variable. An if statement ❷ tests the randomly generated number and draws a water tower on top of the skyscraper only if the number generated is less than 0.5. That means the skyscraper will have a water tower approximately 50 percent of the time, and no water tower the other 50 percent of the time.

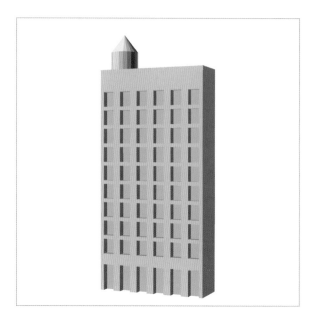

Figure 6-4: A skyscraper with a water tower

Next, let's use random numbers to create a city block of randomly sized skyscrapers (Figure 6-5):

```
❶ use <skyscraper.scad>

   num_buildings = 5;

❷ width_list = rands(10, 30, num_buildings);
   length_list = rands(20, 30, num_buildings);
   height_list = rands(20, 100, num_buildings);

   window_row_list = rands(2.5, 10.49, num_buildings);
   window_col_list = rands(2.5, 10.49, num_buildings);

   watertower_list = rands(0, 1, num_buildings);

   for (n=[0:1:num_buildings-1]) {
❸ width = width_list[n];
   length = length_list[n];
   height = height_list[n];

❹ window_rows = round(window_row_list[n]);
   window_cols = round(window_col_list[n]);

   watertower = round(watertower_list[n]);

   translate([0, n*30, 0]) {
❺ skyscraper(width, length, height, window_rows, window_cols, watertower);
   }
}
```

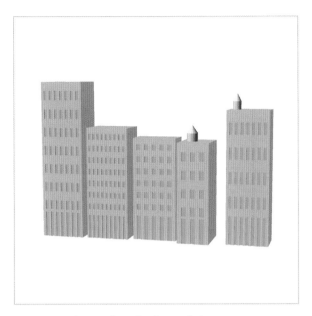

Figure 6-5: A row of randomly sized skyscrapers, some with a water tower

The skyscraper module is imported from *skyscraper.scad* ❶ to keep the design small and manageable. Next, lists of random numbers (of size num _buildings) are generated for each parameter of the skyscraper module ❷. A for loop then draws a number of skyscrapers according to the value indicated by the num_buildings variable. For each new skyscraper, variables are assigned random numbers from the appropriate spot in each list ❸. Decimals are rounded to integer values ❹ for parameters where decimal values wouldn't make sense (you wouldn't want half of a window to be drawn). Finally, this collection of randomly generated values specifies the various parameters ❺ of each new skyscraper. Every time you preview or render this design, each building will be rendered differently, because the random values used to generate each skyscraper will be different. This technique is useful for making repeated computer-generated designs appear more organic and natural.

Summary

This chapter introduced the concept of creating conditional branches with if statements that allow you to create designs that adapt to changing circumstances. Each section of an if statement executes only when a specific condition is true, allowing you to generate designs with varying characteristics. This variety allows you to describe complex designs concisely.

When utilizing if statements to create dynamic designs, keep these concepts in mind:

- if statements use a Boolean expression to evaluate whether a condition is true or false.
- if statements execute only if their Boolean expression is true.
- All expressions in OpenSCAD are evaluated according to an order of operations, which means that a complex Boolean expression can be evaluated from the inside out.
- A nested if statement is an if statement placed inside another if statement.
- To indicate what should happen when a Boolean expression is false, extend an if statement with an else statement.
- You can combine several mutually exclusive decisions in one extended if statement.
- An else statement allows you to provide a default collection of statements that execute when none of the Boolean conditions in an extended if statement are true.
- You can use if statements with random numbers to generate an organic naturalness to your design.
- if statements can help you organize your design into modes (like "print" or "design"), making it easy to change important configuration details.

Before reading further, practice the skills you learned in this chapter by building each of these complex designs (Figure 6-6). Be sure to use at least one if statement in each exercise.

1. Wave

2. Brick wall

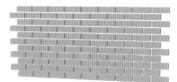

3. Necklace

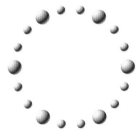

4. Wallpaper

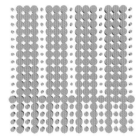

5. LEGO bricks with hole (when possible)

6. Random city block

Figure 6-6: Practice creating all of these designs.

Practice the skills you've learned so far by modeling the following three projects.

RANDOM FOREST

Generate a forest (Figure 6-7) by using random numbers to choose from a variety of tree modules. Design each tree module to take width and height as parameters.

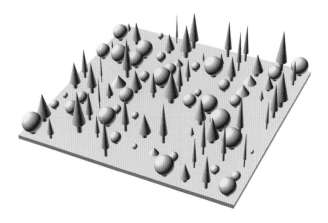

Figure 6-7: Random forest

CLOCK

Design a clock that uses `if` statements to vary the design characteristics of its face markings. Try to emphasize 15-minute positions (Figure 6-8). It may help to use nested `if` statements to apply different transformation operations to various sections of the text-based hour labels of the clock. Applying the same operations to every number won't produce the orientations you see in Figure 6-8.

Figure 6-8: Clock

CITY OF RANDOM SKYSCRAPERS

Extend various exercises in this chapter to generate an entire city of randomly sized skyscrapers (Figure 6-9).

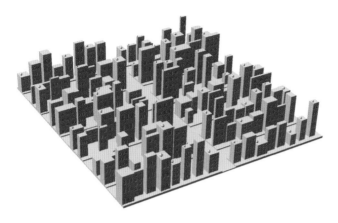

Figure 6-9: City of randomly sized skyscrapers

7

DESIGNING BIG PROJECTS

In this chapter, you'll extend the lessons you've learned so far to build a complex design with OpenSCAD. Specifically, you'll employ an iterative design cycle to plan and complete a larger project. First, you'll apply *computational thinking* to analyze and plan your design. Then, you'll apply the popular *walking skeleton* approach to evolve a low-fidelity prototype from a basic, abstract design into a highly detailed final design. Using this method, you'll connect all the project's major components before fleshing out each component's individual details. As a final step, you'll fill in the smaller details to finish the project.

The Design Cycle

The *design cycle* is a common methodology with four sequential stages to help develop solutions to complex design projects:

Investigate

Understand what you're trying to accomplish. What important considerations or constraints might affect your solution? What do you need in order to accomplish your goals? Can you picture what you're trying to build?

Plan

Divide the process for building your solution into a series of steps. Because you're designing with OpenSCAD (a programming language), you can apply computational thinking concepts (*decomposition*, *abstraction*, *finding patterns*, and *algorithms*) at this stage of the design cycle to identify the best approach to accomplish your goals.

Create

Follow your plan. Creation often reveals new problems, so it's better to build big-picture solutions before focusing on the details. Using a walking skeleton approach to develop a complex design can help make it easier to repeat the Create stage several times. Each repetition of the Create stage (called a design *iteration*) adds more detail to the overall design, allowing you to focus on the most important structural details first.

Evaluate

Compare each iteration of the Create stage (what you've actually built) with the original problem (what you intended to build). Identify areas of concern and then repeat any step of the design cycle as needed.

Keep in mind that the stages of the design cycle are more like a looping cycle. You will probably revisit stages several times throughout the process until you are satisfied with your final design.

Leaning Tower of Pisa Model

Let's follow the design cycle to create a model of Italy's famous Leaning Tower of Pisa (Figure 7-1).

The focus of this project is to combine the design process with computational thinking, so we'll create a recognizable likeness of this famous building, rather than an architecturally accurate scale model.

Figure 7-1: The Leaning Tower of Pisa (photo by Svetlana Tikhonova, covered by the CC0 1.0 Universal [CC0 1.0] Public Domain Dedication license; replicated in Figures 7-2 to 7-4)

Step 1: Investigate—Define Multiple Views

The first step is to search for photos of the Leaning Tower of Pisa to help visualize the final design. We collected images showing different views to provide a sense of what the building looks like from every angle, including front, back, left, right, and top. We (unsurprisingly) couldn't find a photo of the bottom view, but we looked for photos that clearly show how the tower interacts with the ground.

The *Investigate* step of the design cycle is important even if you want to build something of your own invention. If you can't find an exact picture of what you want to build, look for something similar. If you don't have any luck, sketch a rough draft of your intended design by hand. Visualizing your design *before* you code it will save you much time and frustration. The idea is to draw a map of your development process before typing a single line of code.

Step 2: Plan—Apply Computational Thinking

With a firm understanding of what the Leaning Tower of Pisa looks like, you'll analyze the building to identify where you can apply the principles of *computational thinking*: decomposition, patterns, abstraction, and algorithms. Applying these principles when creating designs with OpenSCAD (or any other programming language for that matter) will help you work smarter, not harder, and will allow the computer to do the tedious work for you.

Decomposition

Decomposition is the process of breaking a large, complex problem into smaller, easier-to-describe subproblems, which helps you recognize when to create modules and separate files for a large project. One way to decompose the Leaning Tower of Pisa is to divide the building into three distinct sections (bottom, middle, and top), all of which are "leaning" at the same angle. You then can break those three sections into smaller subcomponents, like columns, levels, fences, and archways (Figure 7-2).

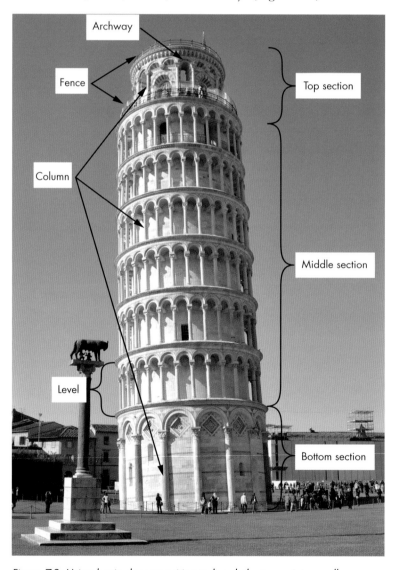

Figure 7-2: Using basic decomposition to break the tower into smaller components

Patterns

Finding *patterns* in a design is a bit like decomposition, because the goal is to break a complex design into smaller, more manageable pieces. However, the objective with patterns is to summarize the process by which elements repeat (Figure 7-3).

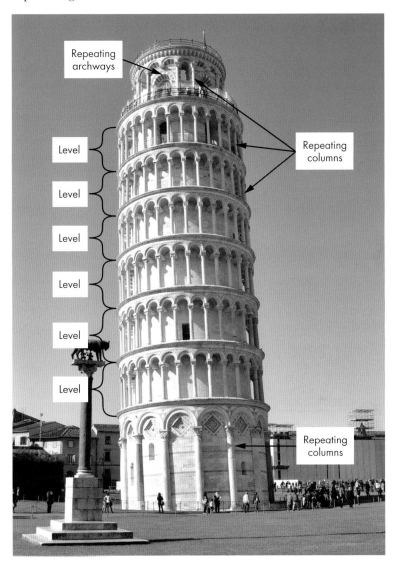

Figure 7-3: Patterns of repeating shapes

For instance, the middle section of the Leaning Tower of Pisa is composed of essentially the same group of shapes repeated six times. Each of those "levels" also includes repeated arches/columns around its outside circumference. In fact, both the bottom and the top sections also contain repeated arches/columns (although at different sizes and intervals from the middle section). Additionally, the top section has two fences with repeated posts, as well as a repeated archway shape in numerous sizes.

Abstraction

Abstraction is the process of summarizing smaller details with higher-level descriptions in order to communicate big-picture information. Rendering each section of the Leaning Tower of Pisa as a cylinder is a general abstraction that omits a lot of detail (Figure 7-4).

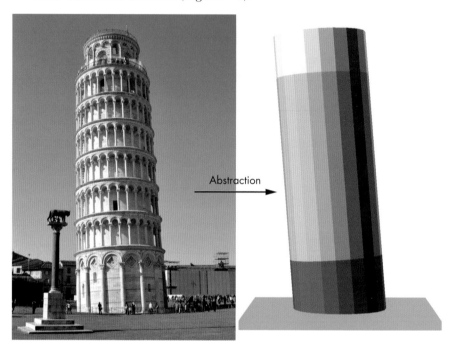

Figure 7-4: Diagram of the Leaning Tower of Pisa abstracted as three cylinders

Abstracting the three sections as cylinders allows you to focus on larger elements (like the angle of the tower's lean and each section's proportional sizing) before considering the smaller, less consequential features.

Algorithms

Because so much repetition exists within the Leaning Tower of Pisa's architecture, our design *algorithm* for creating the tower requires numerous loops. For instance, the columns around the tower's perimeter involve a loop that repeatedly increments the angle of rotation. The looping columns occur in all three sections (bottom, middle, and top), although each section contains different numbers of repeating columns of various sizes.

The multiple use cases for the different sizes of columns around the tower's perimeter suggest that a parameterized column module would be an appropriate algorithmic choice; incorporating parameters in the module allows you to reuse the same basic code for each section of the tower. In fact, the design for this project provides many opportunities to use modules in your code. Each of the basic components you identify during a project's Decomposition and Patterns analysis will likely be a candidate for a module. In this case, you can create modules for the top section, middle section, bottom section, level, column, archway, and fence.

Step 3: Create—Use a Walking Skeleton Approach

The goals of the first two steps of the design cycle are understanding what you want to build and creating a well-defined strategy for breaking a large, complex project into a collection of manageable pieces. In step 3, you start coding by using the walking skeleton development process, allowing you to evolve the design from rough building blocks into a final highly detailed finished piece. You'll use this approach to create several versions of the tower, making incremental improvements with each design iteration (Figure 7-5).

Figure 7-5: Using the walking skeleton approach for the evolution of the Leaning Tower of Pisa

The first versions of the top, middle, and bottom sections in Figure 7-5 are rough abstractions of the final, detailed versions of those same sections. The design's main pieces are connected first as an architectural skeleton, then fleshed out over time in an evolutionary process—hence the name, *walking skeleton*.

Step 4: Evaluate—Decide Which Design Process Steps to Repeat

The "final" step of the design cycle is more of a question than anything else. Does your design accomplish what you intended? Based on the answer, decide which steps of the design process you need to revisit.

To answer that question for the tower example, you'll visually compare the rendered OpenSCAD model of the tower with a photograph of the real Leaning Tower of Pisa. In fact, you'll apply the Evaluate step after each iteration of the walking skeleton to determine which features to add for the next iteration.

Walking Skeleton: Building the Leaning Tower of Pisa

For the remainder of this chapter, you'll build several versions of the Leaning Tower of Pisa in a series of design iterations to demonstrate the walking skeleton development process. Each version will add more details, so you'll compare each iteration with the reference photo and reconsider your plan and algorithms as you go. This approach allows you to apply the design cycle to each iteration without having to worry too much about the way the code is organized or connected.

Iteration 1: Connecting the Tower's Basic Building Blocks

The goal for the first version of the tower design is to create and connect the building's three sections: top, middle, and bottom. You'll also include a platform for stability (the tower is leaning, after all).

Decomposing the building's overall design into smaller pieces provides the setup to evolve the design in stages, as you'll be able to edit the tower's various sections independently. Initially, you'll generate only basic cylinders as big-picture approximations of each section's design, because the first stage of a walking skeleton focuses solely on connecting the project's separate building blocks (Figure 7-6).

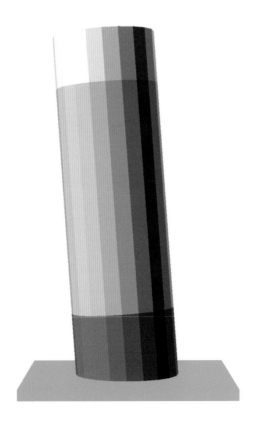

Figure 7-6: An abstract tower with three sections

Although you could use a series of modules contained within one very large file, you'll instead separate these sections into stand-alone files (*bottom.scad*, *middle.scad*, and *top.scad*) and create one connector file (*tower .scad*). Having the code in separate files allows you to create, find, and edit relevant modules for each section easily. You could also use this multi-file approach to collaborate with others, so each person could focus on a different file simultaneously.

The trickiest part of this first step is considering how the different components of the design interact with each other. Usually, this means identifying the crucial information each piece of the design needs in order to be drawn. For instance, to draw an abstract, cylinder-based representation of each section, you need, at minimum, a height and radius for that section. The main project file (*tower.scad*) will communicate that information to each section via module parameters.

Because the top, middle, and bottom sections all use a cylinder as an abstract representation of the final design, creating those files first is relatively easy. The code for each section looks very similar at this stage of the design, which is another advantage of abstraction. You don't need to worry about small details at the moment, so you can copy and paste code in the three files with only minimal changes.

The *bottom.scad* file defines a cylinder to create a simple version of the tower's lowest section:

```
// bottom.scad v1
❶ module bottom_section(width, height) {
    radius = 0.5 * width;
    cylinder(h=height, r=radius);
}
```

The *tower.scad* file communicates the dimensions for the bottom section to the bottom_section module via the width and height parameters ❶.

Next, the *middle.scad* file defines a starting version of the middle section:

```
// middle.scad v1
❶ module middle_section(width, height) {
    radius = 0.5 * width;
    cylinder(h=height, r=radius);
}
```

Again, the *tower.scad* file communicates the width and height to the middle_section module via the width and height parameters ❶.

Similarly, the *top.scad* file defines a basic cylinder to represent the tower's top section:

```
// top.scad v1
❶ module top_section(width, height) {
    radius = 0.5 * width;
  ❷ cylinder(h=height, r=radius);
}
```

As with the bottom and middle sections, the *tower.scad* file uses parameters to supply needed dimensions to the top_section module ❶. The order and number of parameters in each of the three modules is the same. This is a deliberate choice to simplify the design's architecture. As the complexity of the design increases, this consistent interface between *top.scad*, *bottom.scad*, *middle.scad*, and *tower.scad* will make adjusting the proportions of each section easier. The decision to think of each cylinder's measurements in terms of the structure's radius rather than its diameter ❷ was also deliberate (though somewhat arbitrary). At this stage, using width as the cylinder's diameter would also make sense.

Next we create *tower.scad*, which provides the necessary dimensions and connects the tower's three sections with the platform:

```
// tower.scad v1
❶ use <bottom.scad>
use <middle.scad>
use <top.scad>

❷ tower_height = 100;
tower_width = 0.3 * tower_height;
bottom_height = 0.2 * tower_height;
middle_height = 0.65 * tower_height;
top_height = 0.15 * tower_height;

base_width = 2 * tower_width;
base_height = 0.1 * tower_width;

lean_angle = 4;

❸ $fn = 20;

❹ rotate([lean_angle, 0, 0]) {
    color("grey") {
        bottom_section(tower_width, bottom_height);
    }
    color("lightgrey") {
        translate([0, 0, bottom_height])
            middle_section(tower_width, middle_height);
    }
    color("white") {
        translate([0, 0, bottom_height + middle_height])
            ❺ top_section(tower_width, top_height);
    }
}

color("lightgreen") {
    ❻ cube([base_width, base_width, base_height], center=true);
}
```

The first section of the *tower.scad* file links to the three files described previously that define the tower's top, middle, and bottom sections ❶. The next section defines variables to help organize the tower's important characteristics ❷.

Since the design includes not only the tower but also a platform for stability, you create variables to organize the overall tower's height and width (`tower_height` and `tower_width`), the height of each section of the tower (`bottom_height`, `middle_height`, and `top_height`), the height and width of the platform (`base_height` and `base_width`), and the overall angle of the "lean" of the tower (`lean_angle`). You initially set the `tower_height` variable to an arbitrary value, and then use it as part of the definition for most of the other variables. For instance, the height of the bottom section is 20 percent of the `tower_height` variable, so if you want to change the size of the entire design, you need to change only the `tower_height` variable's value.

Next, you use a relatively small number of segments (20) to approximate curved shapes to speed up the rendering of the initial designs ❸. The last design iteration increases the number of segments to 100 in order to generate smoother curved surfaces in the final design.

To avoid duplicating the same rotate operation for all three sections, you use a single operation to apply a consistent angle of rotation to each of the three sections ❹. Each section is called via the appropriate module, with parameters to adjust its width and height. The `translate` operation moves the middle and top sections along the z-axis ❺.

Finally, you draw the platform as a simple cuboid ❻. You also apply different colors to the ground and each section to signify basic proportionality.

From this point on, you won't need to make major changes to the *tower.scad* file. Your initial efforts to size and place each section correctly will form the architectural "skeleton" of the tower design, while your next design iterations will fill in missing details for the tower's top, middle, and bottom sections. The only changes you might need to make to this file in the future would involve adjusting parameters to tweak proportionality as your design evolves, or changing `$fn` to increase the rendered model's smoothness. You'd simply swap out numerical values rather than write new code statements to make those changes.

Iteration 2: Finding Repetition in the Middle Section

Let's take a closer look at the tower's middle section (*middle.scad*) for the second iteration and apply some computational thinking techniques from the planning stage—namely, decomposition and finding patterns. In the middle section, the same collection of shapes (or levels) repeats vertically six times (Figure 7-7).

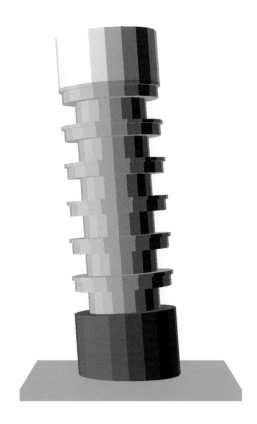

Figure 7-7: Abstract Leaning Tower of Pisa with a looping middle section

Figure 7-8 shows just one of those repeated level shapes.

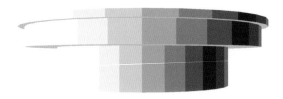

Figure 7-8: A single level shape

To create these repeated levels, you need to make the following changes to the *middle.scad* file:

```
// middle.scad v2
level(50, 25);

module middle_section(width, height) {
    level_height = height / 6;
    level_radius = 0.5 * width;

  ❶ for (h=[0:1:5]) {
      floor_offset = h * level_height;

      translate([0, 0, floor_offset])
        level(level_radius, level_height);
    }
}

❷ module level(level_radius, level_height) {
      lip_height = 0.1 * level_height;
      inner_radius = 0.7 * level_radius;
      overhang_height = 0.3 * level_height;
    ❸ overhang_radius = 0.95 * level_radius;

      // lip
      translate([0, 0, level_height - lip_height])
        cylinder(h=lip_height, r=level_radius);

      // overhang
      translate([0, 0, level_height - lip_height - overhang_height])
        cylinder(h=overhang_height, r=overhang_radius);

      // inner structure
      cylinder(h=level_height, r=inner_radius);
}
```

These changes add more detail to the middle section so it's no longer an abstract cylinder. The level module ❷ organizes all the shapes that construct each floor of the middle section, and a for loop ❶ creates a new level shape repeatedly for each of the six floors in the section. Each level of this section now includes a lip that extends to the full radius of the tower, an overhang that provides a ceiling for columns, and an inner structure to house stairs, doors, and so forth. You create several variables to relate the size of each level feature (lip_height, inner_radius, overhang_height, and overhang_radius) to the level module parameters (level_radius and level_height) ❸.

With this repeating level module, you can simultaneously update all six floors at once by making a change in exactly one place. For instance, if you want to make each level's lip a little thicker or change the overhang radius to provide more room for columns, you can make a single, simple change to the level module definition. Because you are adding detail to only the middle_section module in this phase of our walking skeleton approach, *middle.scad* is the only file you needed to update for the second iteration of the tower design.

To see these new changes reflected in the overall design (Figure 7-7), save *middle.scad*, and then preview the entire design in *tower.scad*. In addition to making your design changes permanent, saving the *middle.scad* file lets OpenSCAD know you want other files to use the updated code. If you want to see the middle_section or level shapes in isolation, create the shape at the top of *middle.scad* and then preview that file. You can include a statement to draw a middle_section or level shape in *middle.scad* without worrying that the shape will also automatically show up in other files. Connecting another file with *middle.scad* with a use directive simply means that module definitions from *middle.scad* will be accessible in *tower.scad*. No drawn shapes from *middle.scad* will be shown unless the connected file uses a module from *middle.scad*.

Iteration 3: Adding More Details to the Middle Section

The next pattern to consider in your computational thinking is the repetition of columns and arches along each floor's perimeter in the middle section (Figure 7-9).

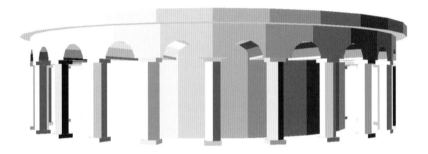

Figure 7-9: A level with repeated columns

To apply these new patterns to the design, you create a column shape and repeat that new shape along the circumference of the level module. This means you need to modify the *middle.scad* file again, as that's where the level module is defined. To create a column shape, you also define a column module in a new *column.scad* file.

In the design cycle's planning phase, you noticed that columns and arches repeat around the circumference of each of the tower's three sections. Because you need to include column shapes in multiple files, defining the column module in a separate file makes it easier for different sections to use that new shape definition. Columns and arches repeat in different patterns in each section, and they also vary in their ornamentation. That's why at this initial stage, you'll focus on creating an abstract column with basic components (Figure 7-10). You can then update this basic definition of a column in a later design iteration.

Figure 7-10: An abstract column

Creating a column module in a separate file called *column.scad* makes it easier to share and evolve your use of columns in the future as needed:

```
// column.scad v3
❶ module column(col_width, col_height) {
    col_radius = 0.5 * col_width;
  ❷ orn_height = 0.05 * col_height;

    translate([-col_radius, -col_radius, col_height - orn_height])
        cube([col_width, col_width, orn_height]);
    cylinder(h=col_height, r=col_radius);
    translate([-col_radius, -col_radius, 0])
        cube([col_width, col_width, orn_height]);
}
```

As with other modules, you include two parameters (`col_width` and `col_height`) in the `column` module ❶ to provide the necessary information to create a column shape. Based on the column height and column width, variables are created (`col_radius` and `orn_height`) to describe the column's radius and the ornamentation's height included at both the top and bottom of a column ❷. While it may seem to make the module definition more complicated, defining and using these variables rather than placing repeated arithmetic calculations as module parameters or inside operations reduces the number of possibilities for error, groups all of the design assumptions at the top of the module, and makes it easier to update all uses of a measurement.

To invoke this new `column` module, you then modify the `level` module in *middle.scad* to draw repeating columns and arches around the circumference of each level:

```
// middle.scad v3
❶ use <column.scad>
...
module level(level_radius, level_height) {
  ❷ lip_height = 0.1 * level_height;
    inner_radius = 0.7 * level_radius;
    overhang_height = 0.3 * level_height;
    overhang_radius = 0.95 * level_radius;

    num_cols = 24;
    angle_size = 360 / num_cols;

    col_height = 0.65 * level_height;
    col_width = 0.2 * col_height;

    arch_depth = 2 * (level_radius - inner_radius);

    // lip
    translate([0, 0, level_height - lip_height])
        cylinder(h=lip_height, r=level_radius);

    translate([0, 0, col_height]) {
        difference() {
            // overhang
            cylinder(h=overhang_height, r=overhang_radius);

            // arches
          ❸ for (i=[0:1:num_cols-1]) {
                angle = i * angle_size + angle_size/2;
                rotate([0, 0, angle])
                    translate([inner_radius, 0, 0])
                        rotate([0, 90, 0])
                            cylinder(h=arch_depth, r=col_width, center=true);
            }
        }
    }
}
```

```
    // inner structure
    cylinder(h=level_height, r=inner_radius);

    // columns
❹ for (i=[0:1:num_cols-1]) {
        angle = i * angle_size;
        rotate([0, 0, angle])
            translate([overhang_radius - 0.5 * col_width, 0, 0])
                column(col_width, col_height);
    }
}
```

Comparing this updated version of *middle.scad* with the version from your second design iteration reveals three major additions to the level module. First, *column.scad* is connected to this file ❶ with a use directive so that you can use the new column module to draw column shapes in this file. Next, variables are defined to describe the number of columns per level (num_cols), the angle at which the columns should be repeated along the circumference of the tower (angle_size), the width and height of each column (col_width and col_height), and the depth of the arch connecting every two columns that will be carved away from the overhang of each level (arch_depth) ❷.

After creating the overhang, you include a for loop within a difference operation to carve away arches between the location of each column ❸. A final for loop repeats columns along the level's circumference ❹. You could combine these two loops into a single for loop that uses an if statement; however, the loops are separated here to make the logic clearer.

As before, to see these new changes reflected in the overall design, save both *middle.scad* and *column.scad*; then preview the entire tower design in *tower.scad*. To see only the middle section without the rest of the tower, include a statement to draw a middle_section shape at the top of *middle.scad*; then preview the design in *middle.scad*. You can also easily see only a column shape by including a statement to draw a column shape at the top of *column.scad* and then previewing the design in that file.

After using a relatively small amount of code to add a large number of repeating columns and arches to the middle section, that section of the tower (Figure 7-11) is now more recognizably similar to our reference photo of the Leaning Tower of Pisa (Figure 7-1).

However, as you can see in Figure 7-11, the top and bottom sections are still abstract simplifications. Applying the design cycle's Evaluate step after each iteration of the walking skeleton helps identify missing details that might offer the most noticeable improvements to a design. After this iteration, you should once again consult the reference photo (Figure 7-1) to decide which section of the tower now most needs improvement.

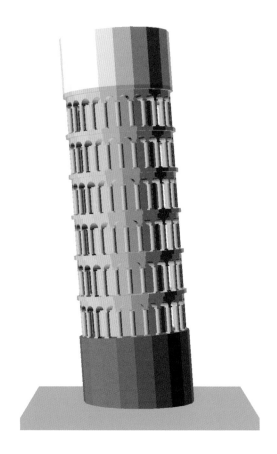

Figure 7-11: Leaning Tower of Pisa with modularized columns

Iteration 4: Adding Details to the Top Section

The top of the tower is missing fences, repeating columns, and archways (windows and doors), so the next iteration focuses on adding those details. You'll add two fences to the top section, as well as alternating archways of different sizes and heights (Figure 7-12), so you'll modify *top.scad* by adding a fence module and an archway module. You'll draw the archway module in different sizes to create the doors and windows shown in the top section of our reference photograph (Figure 7-1).

Figure 7-12: Fenced-in top section with alternating archways of different sizes

This updated version of the *top.scad* file adds the fence and archway details to the tower's top section:

```
// top.scad v4
module top_section(width, height) {
❶ top_radius = 0.4 * width;
   room_radius = 0.75 * top_radius;

   num_doors= 5;
   door_angle= 360 / num_doors;

   overhang_height = 0.1 * height;
   overhang_width = 1.1 * top_radius;

   door_height = 0.6 * height;
   door_width = 0.35 * height;

   window_height = 0.25 * height;
   window_width = 0.15 * height;
```

```
    // overhang
    translate([0, 0, height - overhang_height])
        cylinder(h=overhang_height, r=overhang_width);

    //inner structure
    difference() {
        cylinder(h=height, r=top_radius);

        translate([0, 0, 1]) {
            cylinder(h=height-2, r=room_radius);

          ❷ for (i=[0:1:num_doors-1]) {
                angle = i * door_angle;
                rotate([0, 0, angle])
                    translate([top_radius-2, 0, 0.25*height])
                        // doors
                        archway(door_height, door_width, room_radius);
                rotate([0, 0, angle+0.5*door_angle])
                    translate([top_radius - 2, 0, 0.6*height])
                        // windows
                        archway(window_height, window_width, room_radius);
            }
        }
    }

    //fencing
    translate([0, 0, height])
        fence(15, 3, top_radius, 1);
  ❸ fence(20, 3, 0.5*width, 1);
}

❹ module fence(num_posts, fence_height, fence_radius, post_width) {
    post_radius = 0.5 * post_width;
    angle_size = 360/num_posts;
    ring_height = 0.5;
    post_height = fence_height - ring_height;

    translate([0, 0, post_height])
        ring(fence_radius - post_width, fence_radius, ring_height);
    translate([0, 0, post_height / 2])
        ring(fence_radius - post_width, fence_radius, ring_height);

    for (i=[0:1:num_posts-1]) {
        angle = i * angle_size;
        rotate([0, 0, angle])
            translate([fence_radius - post_radius, 0, 0])
                cylinder(h=post_height, r=post_radius);
    }
}

❺ module ring(inner_radius, outer_radius, height) {
    difference() {
        cylinder(h=height, r=outer_radius);
        translate([0, 0, 1])
```

```
            cylinder(h=height+2, r=inner_radius, center=true);
        }
    }

❻ module archway(height, width, depth) {
        radius = 0.5 * width;

        rotate([90, 0, -90]) {
            translate([0, (height - radius) / 2, -depth / 2])
                cylinder(h=depth, r=radius);
            cube([width, height - radius, depth], center=true);
        }
    }
```

As with the other module definitions, you begin by defining variables to describe the top section's various features ❶. You base the number of windows on the number of doors (num_doors), but otherwise, you deliberately choose variable names that are self-documenting. A for loop contained within a difference operation subtracts repeated windows and doors from the top section's inner structure ❷. Windows and doors have similar shapes, so you define a single archway module that lets you vary the size of window and door shapes with the height, width, and depth parameters ❻.

The top_section module ends by drawing two fence shapes ❸. These fences are basically the same shape but different sizes, so you define a fence module to construct them ❹. You also include a ring module to make it easier to create various fencing rings ❺. This definition of a ring module is transferred from a previous Design Time activity (see Chapter 5). Reusing modules from prior projects can save a lot of time and effort.

To simplify the project's organization, you include the fence, ring, and archway modules only in the *top.scad* file since no other section contains those shapes. As with previous design iterations, save your updates to *top.scad*; then preview the design to see those changes in other files.

The top_section module now produces a more detailed version of the top of the tower (Figure 7-13).

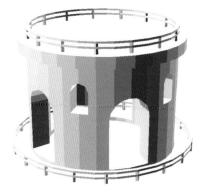

Figure 7-13: Fenced-in top section with alternating archways, detail view

Comparing this design iteration with the tower's reference photo (Figure 7-1), your evaluation suggests that the bottom section now needs the most attention.

Iteration 5: Adding Details to the Bottom Section

This update modifies the *bottom.scad* file to include the major missing features (columns and arches):

```
// bottom.scad v5

❶ use <column.scad>

module bottom_section(width, height) {
    radius = 0.5 * width;
    inner_radius = 0.9 * radius;
    lip_radius = 1.05 * radius;
    lip_height = 0.05 * height;
    overhang_height = 0.2 * height;

    num_cols = 14;
    angle_size = 360 / num_cols;
    col_height = height - overhang_height;
    col_width = 0.1 * col_height;

    // lip
    translate([0, 0, height - lip_height])
        cylinder(h=lip_height, r=lip_radius);

    // inner structure
    cylinder(h=height, r=inner_radius);

    // columns
❷ for (i=[0:1:num_cols-1]) {
        angle = i * angle_size;
        rotate([0, 0, angle])
            translate([radius - 0.5*col_width, 0, 0])
                column(col_width, col_height);
    }

    // arches
    translate([0, 0, col_height])
        difference( ) {
          // overhang
          cylinder(h=overhang_height, r=radius);

          // arches
        ❸ for (i=[0:1:num_cols-1]) {
            angle = i * angle_size + angle_size/2;
            rotate([0, 0, angle])
                translate([inner_radius, 0, 0])
                    rotate([0, 90, 0])
```

```
                    cylinder(h=radius-inner_radius, r=col_width);
        }
    }
}
```

You first include *column.scad* in order to access the `column` module ❶. This allows you to use a `for` loop to draw columns around the bottom section's perimeter ❷. Columns in the bottom section are bigger than those in the middle section, so parameters for drawing a column are set accordingly. You add the arches next, also with a `for` loop ❸.

Save *bottom.scad* and then preview the design to reveal new details in the tower's bottom section (Figure 7-14).

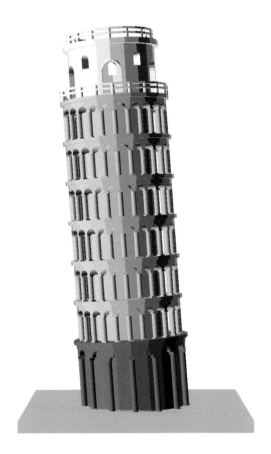

Figure 7-14: Tower with the updated bottom section

The tower is now visually similar to the actual Leaning Tower of Pisa. You can apply the Evaluate stage one more time, but adding more details might not produce much benefit if you intend to make a small 3D print of the model.

Final Evaluation of the Design Cycle

At this stage, the tower looks very similar to the Leaning Tower of Pisa. Making a slight modification to $fn in *tower.scad* increases the design's smoothness, providing an even closer likeness (Figure 7-15).

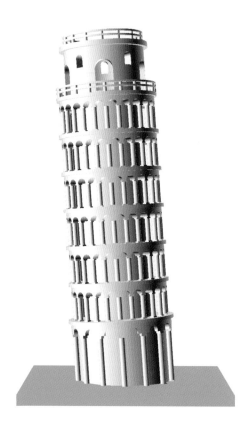

Figure 7-15: Smoother tower with $fn=100 instead of $fn=20

You left the smallest details for last, which is a deliberate feature of the walking skeleton approach to project development. Every design iteration focuses on one major area, specifically chosen to provide the most noticeable improvement to the overall tower design. As mentioned previously, because you plan to 3D-print this model, you omit especially small details, but could have included the following:

- The missing columns and arches from the top section.
- The missing rectangular doorways from the middle and bottom sections.
- The different ornamentation of columns and arches in each section.
- Columns are not basic cylinders, so you could have given the top of a column a smaller radius than the bottom.

We mention these missing features as potential exercises for readers who want to continue doing design iterations of this model. Larger 3D prints potentially could reveal those smaller design features.

Design Organization Overview

For your first design iteration, you split the building into three low-fidelity sections, each having a separate *.scad* file. This way, all you needed to do was preview only one file (*tower.scad*), because that file connected together the three other files. Figure 7-16 shows the initial project's organization, which reduced the amount of code in any one file, making it easier to find and modify specific parts.

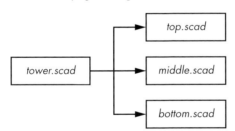

Figure 7-16: Initial architecture for the Leaning Tower of Pisa project

Throughout the design process, you used decomposition to find opportunities to break larger components of the tower into smaller pieces. After your last iteration, the project organization evolved to contain many modules and an additional file (Figure 7-17). This final project organization illustrates the main principle of the walking skeleton approach to development. Your initial project organization focused on connecting big pieces of the project, while your final organization reveals all of the smaller details you added incrementally during each iteration.

The organization and development process described here is only one way to build this project. Aside from organizing the project into a different collection of separate *.scad* files (or even one massive *.scad* file), you could have created a different set of modules to decompose the tower into smaller building blocks.

We also missed several opportunities to reduce the need for repeating code by including additional if statements or for loops. For instance, you could have created a separate column_ring module to "factor out" the inclusion of columns and arches around the tower's circumference. With careful use of if statements and parameters, you could have used the column_ring module to draw both the columns and arches in all three sections, greatly simplifying the code required in the top_section, middle_section, and bottom _section modules.

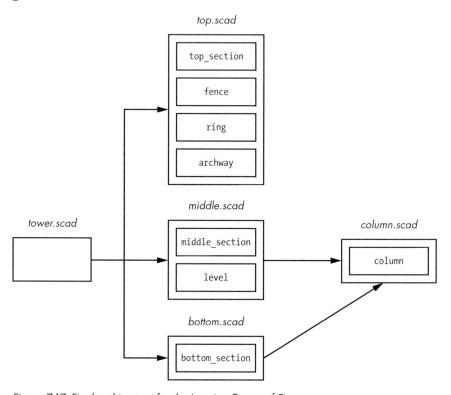

Figure 7-17: Final architecture for the Leaning Tower of Pisa

A design can evolve over time without major changes to the overall project's organization. You don't need to know all the modules or files you'll need to create at the beginning of a project; you can make those decisions as you gain a better understanding of what you're building. Each time you apply the Evaluate stage of the design cycle, you have an opportunity to reconsider which changes to make to your design.

Summary

This chapter introduced the benefits of deliberately following the design cycle when building a complex project. You applied computational thinking to guide the planning phase and a walking skeleton approach to combine the Build and Evaluate stages into a looping procedure. You connected the design's most important features first and then incrementally developed each component's major features. Only during the final stages of development did you consider the smaller, more nuanced details.

To recap, keep these concepts in mind when designing a complex project:

- Draw a sketch of the project you want to build, and label it with patterns, abstractions, and decompositions to help you understand how to organize your code.
- Describing the minimum information needed to draw a new shape can help guide you to understand which parameters might be necessary for a new module.
- Using self-documenting naming conventions will help organize your code by revealing the purpose of each new variable or module.
- Use color to help organize different pieces of an evolving design.
- Make sure to save individual files when you make any changes, so other files can use the newest version of that file.
- Connect your project's most important pieces first, even if those pieces are big-picture abstractions.
- Design a project's smallest details in the final stages of your walking skeleton development approach.

The design cycle and walking skeleton development model are common approaches, and you can find abundant material online for further reading. We encourage you to explore these concepts further as you create new designs with OpenSCAD.

AFTERWORD

After reading this book, especially if you engaged with the Design Time and Big Project exercises, you should have a solid grasp on how to create 3D-printable designs with the OpenSCAD programming language. In closing, we'll provide some helpful pointers on where to go next, as well as help contextualize how OpenSCAD fits into the larger ecosystems of the open source and maker movements.

Learn More About OpenSCAD

We have covered a significant portion of OpenSCAD's available features here; however, there are still more advanced features to uncover. A variety of resources are available for you to unlock OpenSCAD's full creative power:

Visit the OpenSCAD online documentation

Once you're ready to take your OpenSCAD skills further, your first stop should be the official online OpenSCAD documentation (*https://openscad.org/documentation*). This is the place to look for other well-structured guides to learning more about OpenSCAD. You'll find tutorials, a user manual, a more complete language reference, and regularly updated links to many other learning materials, including books, articles, and videos.

Remix someone else's OpenSCAD design

For slightly less structured learning, try to remix an existing OpenSCAD design. Learning to read code written by other people can result in substantial improvements to your own coding and organizational skills. You can easily search online for OpenSCAD designs and be sure to check out two of the most popular 3D design-sharing websites: *https://thingiverse.com/* and *https://youmagine.com/*. Searching for *openscad* will result in thousands of OpenSCAD designs available for you to use and remix.

Most of the designs are also available as OpenSCAD code, which you can explore to see how other people solve challenging design problems in code. Creating a remix of someone else's design by integrating your own customized innovations into their code is a great way to demonstrate that you truly understand how all the pieces of their design fit together.

Join the OpenSCAD community

Engaging with other like-minded people in the thriving OpenSCAD community of designers is another way to supplement your learning. Sometimes your design ideas might present unique challenges that no amount of reading or searching will illuminate. Asking the OpenSCAD community for help could offer the perfect solution.

The official OpenSCAD community page (*https://openscad.org/community*) has a chat room as well as a mailing list and forum where OpenSCAD users discuss projects, ask for help, and even facilitate development of OpenSCAD itself. OpenSCAD is open source software, and development discussions often take place in the same forums where community members hang out. In addition to finding answers to your most perplexing design problems, participating in the OpenSCAD forum means that you can offer help to others, and you might even be able to influence the development of the OpenSCAD software itself by suggesting new features or reporting bugs.

The Open Source Ethos

As we've mentioned several times throughout the book, OpenSCAD is open source software. Proprietary 3D design software is typically expensive and usually carries a steep learning curve. Even "free" web-based 3D-design tools often require creating an account, which can raise concerns about privacy or longevity of the service. The OpenSCAD community of developers wanted to create a truly free and accessible 3D-modeling platform to open the world of solid CAD modeling to everyone, especially people interested in the intersection of coding and 3D design. Hundreds of people have donated their time and effort to create and improve OpenSCAD for you, in the hopes that removing some of these traditional barriers will encourage more people to learn and use 3D modeling to solve problems both big and small.

Motivation and Ecosystem

Why would so many people spend so much time and effort to turn something that is traditionally "hard" and "expensive" into something that is both free and so much more accessible and approachable? Important motivating reasons behind making OpenSCAD open source include the following:

- Supporting communities that celebrate cross-cultural and cross-discipline explorations
- Supporting and engaging with the inclusive teaching, learning, and sharing of important STEM/STEAM skills
- Encouraging individuals to share the benefits of their work and efforts with others
- Empowering individuals to make things better by providing change-making ownership of existing creations through a crowd-based iterative design process
- Believing that paying it forward encourages others also to pay it forward, resulting in a magnified benefit to society

In fact, the OpenSCAD open source project also exists because of the kindness of strangers. The OpenSCAD development community relies upon many other open source projects that were each created so that others could use the technology to (hopefully) make the world a better place. Some of the most prominent are as follows:

- Qt to help build the OpenSCAD user interface (*https://qt.io/*)
- CGAL for help evaluating constructive solid geometry (CSG) when OpenSCAD designs are rendered (*https://cgal.org/*)
- OpenCSG and OpenGL to help generate CSG previews for OpenSCAD designs (*http://opencsg.org/* and *https://www.opengl.org/*)
- Boost for its large toolbox of C++ convenience libraries (*https://boost.org/*)
- Eigen to provide fast and well-tested linear algebra functions (*https://eigen.tuxfamily.org/*)

We would like to thank the developers of OpenSCAD and every open source project for their time and valuable contributions.

Online Citizenship

It can be easy to forget that real people are on the other side of the screen you use to access the internet. The open source software movement relies heavily upon the idea of online citizenship, making sure that the distributed social network of the internet helps promote positive social change while supporting the advancement of human rights. Here are a few starting principles for online citizenship we hope you'll take with you as you continue your journey with OpenSCAD and other open source software projects:

Give credit
> Provide attribution when you use something someone else has created. This helps support the original creator (even with kudos) and demonstrates that you're aware of the privileges of "standing on the shoulders of giants."

Have empathy for others
> Remember that the people you interact with online don't necessarily share your background, language, culture, or inside jokes. Maintain and model a respectful and considerate use of communication in all community spaces. Be respectful of the cultural and environmental impact of the things you create.

Pay it forward
> Create things that help solve problems for real people. Share your creations, especially when you've created something by using tools that other people have given away for free.

OpenSCAD and the Maker Movement

It would be an oversight to overlook OpenSCAD's relationship with the maker movement. *Making* has become an increasingly popular term to describe taking a creative, DIY approach to problem-solving. Making usually involves trying to solve a problem by using an iterative design process and a variety of machines, tools, and materials: cardboard prototyping, 3D printing, laser cutting, electronics, soldering, woodworking, sewing, CNC (computer numeric control) machining, vinyl cutting, screen printing, water-jet cutting, and so on.

OpenSCAD is a key software tool for the maker community. Although this book focuses on designing with OpenSCAD in anticipation of 3D printing, 3D printing only scratches the surface of what the maker community has created with OpenSCAD. Combining OpenSCAD and 3D printing is a great solution for many problems, but it's not always the best solution. Developing a far-reaching, holistic sense of the design tools and paradigms collected under the maker umbrella provides many benefits.

Making and Creative Problem-Solving

We have used the word *design* intentionally to describe OpenSCAD creations, because each OpenSCAD project is created for a specific reason, often to solve a physical problem in the real world. Fundamental to the notion of design is the practice of *problem-solving*. Similar to swimming, problem-solving through design is a skill that can be truly learned only when you are "in the water." Every time you finish an OpenSCAD project, you increase your capacity to design a specific solution to a specific problem.

The maker movement rightly recognizes creative problem-solving through design as a transferable skill. If you are new to the maker movement, you might find it surprising that designing a sewing kit can help with your ability to create a well-ordered sequence of OpenSCAD code or that creating a multilayered screen print can help you decompose a complicated problem into well-defined smaller parts. You can acquire these higher-order design skills in any medium. In addition to the transferable programming and 3D-printing techniques you've learned in this book, we hope that you will consider applying your new problem-solving and design proficiency in a few other interesting directions.

2D Fabrication

The world of 2D fabrication is a vast landscape for applying the skills you've learned in this book. Extruding a 2D shadow to create a 3D design is a powerful 3D design tool. However, many maker tools use 2D files (such as *.svg* or *.dxf*) to manufacture physical versions of their design. 2D fabrication machines (such as laser cutters, vinyl cutters, water-jet cutters, and so on) essentially cut the outline of the 2D shape into flat pieces of wood, metal, vinyl, felt, cardboard, or most other flat materials. Because OpenSCAD makes it so easy to use variables, arithmetic, loops, and if statements to place and combine shapes, many makers use OpenSCAD to create purely 2D designs specifically for these machines.

Here are a few ideas to inspire your 2D creativity with OpenSCAD:

- Use a collection of loops to generate small, circular holes along the perimeter of a 2D leather sewing pattern you've created in OpenSCAD. Then, cut out the pattern with a laser cutter or a cutting machine. Leather is difficult to punch a needle through, but using OpenSCAD loops to generate the holes will help save time and effort.

- Use a CNC wood cutter to cut out a life-size version of a piece of flat-pack furniture you've designed with OpenSCAD 2D shapes. Although 3D printers have a relatively small printing area, CNC cutters can cut a rather large surface area. 3D printers can be used for prototyping, while the usable furniture is created on a large CNC machine.

- After you've 3D-printed a few prototype versions of a flat 2D gear you've designed and extruded with OpenSCAD, use a water-jet cutter to cut it out of metal. Plastic gears don't last nearly as long as metal gears, especially if you're actually using them for your bike.

Physical Computing

Many interactive maker projects combine electronics and computers with other physical components to create something with dynamic characteristics. What if your OpenSCAD designs could sense and respond to the world, or even move? A variety of inexpensive, pocket-size computers are available that can supercharge the interactivity of the designs you create with OpenSCAD.

These miniature computing platforms utilize a variety of sensors and output (like microphones, temperature sensors, movement sensors, speakers, motors, and LEDs) to interact with the real world. A few of the most popular small computing platforms are listed here:

- Raspberry Pi (*https://raspberrypi.org/*)
- Arduino (*https://arduino.cc/*)
- micro:bit (*https://microbit.org/*)
- Circuit Playground (*https://learn.adafruit.com/introducing-circuit-playground/*)

Each of these devices has a large online community with plenty of learning resources available. Combine OpenSCAD with one of these inexpensive pocket-size computers to explore areas like robotics, physical computing, wearable computing, human-robot interaction, or the Internet of Things. Here are some examples of projects you might create with OpenSCAD and one of the preceding devices:

- Automatic plant/garden watering systems
- Totally new, interactive digital instruments
- Physical enclosures for personal, multinode data centers
- Personal assistive devices to help with accessibility for people with disabilities

By using OpenSCAD to invent creative new uses for these electronic devices, you're setting up your project to be easily customizable, shareable, and extensible. Maybe you can even create something that kick-starts your own open source project.

Makerspaces

Making can happen anywhere, but getting together at a central location so that like-minded creators can share ideas and troubleshoot solutions has become popular for makers. A *makerspace* is a physical location and community of makers that provides a collection of tools, machines, and learning resources. Makerspaces provide access to equipment that might be too expensive to own personally and serve as a physical community for makers that can offer the same benefits as the virtual communities mentioned previously. You can find makerspaces (big and small, free and fee-based) at libraries, schools, independent venues, and maker faires or festivals (*https://makerfaire.com/*).

If your local community doesn't yet provide access to a centralized, shared location for making, some makerspaces have gone virtual. Many online vendors allow you to upload designs for 3D printing or 2D cutting in a variety of materials, providing an affordable stepping-stone for creating a physical version of your design when you don't have access to a fabrication machine.

Final Ideas for More Practice

We'd like to leave you with some final advice. The key to mastering any skill is a combination of learning and doing. If you have only read this book without actually coding or designing, you've skipped a big portion of the learning potential, but it's not too late! You can put this book down right now and go back to any project.

If you're looking for more examples of well-defined design exercises, the following screenshots present a few visual ideas from the OpenSCAD community that should work well as inspiration for "next step" projects. You can also check out *https://openscad.org/gallery* for more curated examples.

Customizable Measuring Spoons

Creating a measuring spoon module is a good intermediate design project (Figure 1). The primary challenge with this project is to create a single OpenSCAD module with the spoon size, units, and configuration (nested stacking spoons or spoons that lay flat) as parameters. The measuring spoon shape and labels can then be generated from those parameters. Can you generate 3D-printed measuring spoons with the exact measurements required to trust when cooking or baking?

Figure 1: A collection of measuring spoons with several sizes and configurations

These measuring spoons were generated from OpenSCAD code originally designed by charliearmorycom. You can find the Customizable Measuring Spoon project at *https://www.thingiverse.com/thing:51874/*.

Customizable Vacuum Tool

Designing a custom tip to fit the end of a vacuum hose is a good example of a project that needs to interface with an existing physical tool. Creating a well-fit physical connection will require both careful measurement and continued experimentation with a 3D printer in order to perfect dimensions. Additionally, this project offers the opportunity to create a customizable nozzle, which can be generated by one or more module parameters (Figure 2).

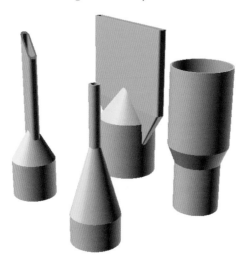

Figure 2: A collection of vacuum tool tips with parameterized nozzles

The OpenSCAD code used to generate these customized vacuum nozzle tips was originally designed by Ziv Botzer. The Customizable Vacuum Tool project can be found online at *https://www.thingiverse.com/thing:1571860/*.

Customizable Flowerpots

Using OpenSCAD to create a flowerpot module will allow you to design something that combines the decorative with the functional. This intermediate-level project will allow you to scale your 3D prints both large and small, depending on the size of the plant you'd like to house (Figure 3). There are several opportunities for parameters in this project, with a bonus challenge of generating both the flowerpot and the saucer tray from the same module. Don't forget to include a hole in the bottom of the flowerpot for water to drain into the saucer!

The OpenSCAD code used to generate this collection of flowerpots came from the Customizable Flower Pot (classic style) project by Robert Wallace, which is available online at *https://www.thingiverse.com/thing:2806583/*.

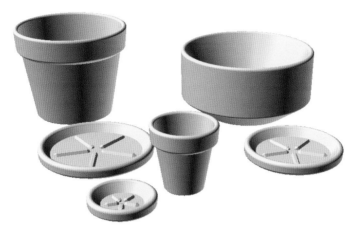

Figure 3: A collection of flowerpots and saucers trays of different sizes and shapes

Drawer Box

Remember the desktop organizer you created as a Big Project in Chapter 2. This box and drawer organizer is a more complex organizer idea, which can be taken in many directions. The initial challenge is to tune the box dimensions in order to make the drawers slide easily, yet firmly, into the box. Customizing sizes and designs of the box, drawers, and drawer layout (ideally via parameterized modules) are also good future challenges. Notice how this project includes useful details like small, spherical nubs on box dividers to keep drawers in place, as well as for loop-generated holes on all three sides of the box to reduce both the time and material required to 3D-print a Drawer Box (Figure 4).

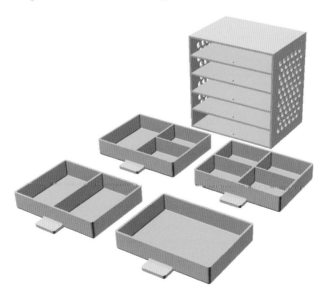

Figure 4: A desktop organizer with several different drawer configurations

The Drawer Box project by Gian Pablo Villamil can be found online at *https://www.thingiverse.com/thing:421886/*.

Lab Clamps

Designed for use in a physics classroom, this project is a good example of using 3D printing to manufacture replacements for items that are normally cost prohibitive. 3D-printing mechanical parts designed to fit with existing tools or parts is always challenging. In this example, a series of clamps and stands are designed to be mated with metal bolts (Figure 5). Designing an appropriate inner structure to mate firmly with these bolts can require some experimentation. Projects like this are a good example of how OpenSCAD and 3D printing can work together as a service project for a school or community center.

Figure 5: A collection of clamps and stands for physics experiments

The Lab Clamps project was created by Mark Schober. You can find the code used to generate the clamps in this picture (along with more details on how to incorporate metal bolts and mass manufacture these parts with silicone molds) at *https://www.modelingscience.org/post/3d-print-your-own-lab-clamps/*.

Chess Set

Designing a chess set is a favorite project among both artists and 3D-printing enthusiasts. While the example shown is very close to a classic chess set (Figure 6) and would likely require sourcing a 3D model of a horse's head, many designs exist online for creating a more contemporary or abstract chess set. Creating a base module would help provide a consistent size and design for your own chess set, while creating a separate module for each piece would make it easy to organize your 3D printing.

Figure 6: A custom chess set

The code used to generate this chess set was designed by Tim Edwards and is available at *https://www.thingiverse.com/thing:585218/.*

Pegboard Wizard

Have you ever needed to organize a collection of tools or hardware using a pegboard? This last example leverages the modular potential of a standard pegboard to create a library of useful container bins and tool holders (Figure 7). Create a single module with many parameters, or a collection of modules with fewer parameters. Either way, this project will test your ability to apply principles of computational thinking while you also create a useful organizational solution to your offline toolkit.

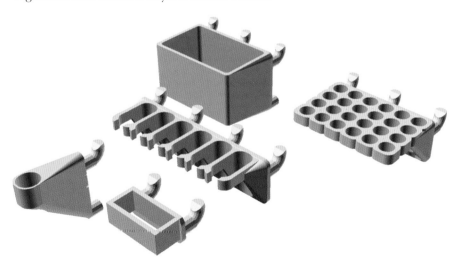

Figure 7: A collection of pegboard organizers created by the pegboard wizard

Pegstr - Pegboard Wizard was designed by Marius Gheorghescu and is available at *https://www.thingiverse.com/thing:537516/.*

A

OPENSCAD LANGUAGE REFERENCE

This language reference provides short descriptions of most OpenSCAD features, serving as a quick reminder of how to use functionality described in this book or a way of discovering new OpenSCAD features. Consult the official OpenSCAD reference at *https://openscad.org/ documentation* for the full manual.

Syntax

Create a 2D or 3D shape with a collection of parameters. Terminate the command with a semicolon (;):

```
shape(...);
```

Create a shape that has been transformed by a series of operations. Terminate the statement with a semicolon (;):

```
transformation2(...) transformation1(...) shape(...);
```

Create a variable to name and refer to an important value; values are assigned once and cannot change:

```
var_name = value;
```

Create a user-defined shape called name with zero or more parameters. User-defined shapes work the same way as built-in shapes:

```
module name(...) { ... }
name(...);
```

Create a user-defined mathematical operation called name with zero or more parameters:

```
function name(...) = ...;
name(...);
or
name = function(...) ...;
name(...);
```

Import and immediately execute the OpenSCAD code in *filename.scad*:

```
include <filename.scad>
```

Import and make usable (but don't immediately execute) the OpenSCAD functions and modules in *filename.scad*:

```
use <filename.scad>
```

Operators

Operators are listed in decreasing order of precedence. When multiple operators from the same level of precedence occur in an expression, the operators are evaluated in order of occurrence (from left to right):

```
^

*, /, %

+, -

<, >, <=, >=

==, !=

&&

||
```

2D Shapes

Draw a circle of the defined radius or diameter:

```
circle(radius | d=diameter)
```

Draw a square with length = *size* and width = *size* (equal sides); optionally center the square at (0,0):

```
square(size, center)
```

Draw a rectangle with width along the x-axis and length/depth along the y-axis defined by a vector; optionally center the square at (0,0):

```
square([width, height], center)
```

Draw a polygon that connects all of the points defined by the vector of [x, y] points:

```
polygon([[x1, y2], [x2, y2], ..., [xn, yn]])
```

Draw a polygon that connects all of the points defined by the vector of [x, y] points; optionally define a collection of paths for polygons with holes:

```
polygon([points], [paths])
```

Draw words defined by the *text* string; optionally specify the size, font, horizontal alignment, vertical alignment, letter spacing, direction, language, and script of the text:

```
text(text, size, font, halign, valign,
spacing, direction, language, script)
```

Import a 2D SVG or DXF file:

```
import("filename.svg")
```

3D Shapes

Draw a sphere centered at (0, 0, 0) with the specified radius or diameter:

```
sphere(radius | d=diameter)
```

Draw a cube with length = *size*, width = *size*, and height = *size* (equal sides); optionally center the cube at (0,0,0):

```
cube(size, center)
```

Draw a cuboid with width along the x-axis, length/depth along the y-axis, and height along the z-axis defined by a vector; optionally center the cube at (0,0,0):

```
cube([width, depth, height], center)
```

Draw a cylinder with the specified height and radius or diameter; optionally center the cylinder at (0,0,0):

```
cylinder(h, r|d, center)
```

Draw a cone with the specified height and radii or diameters; optionally center the cone at (0,0,0):

```
cylinder(h, r1|d1, r2|d2, center)
```

Draw a 3D solid defined by vectors of points and faces; optionally use convexity to improve the preview of complex concave shapes:

```
polyhedron([points], [faces], convexity)
```

Import an STL, OFF, 3MF, or AMF file:

```
import("filename.stl")
```

Draw a 3D height map of the data file; optionally center the shape at (0,0) and use convexity to improve the preview of complex concave shapes:

```
surface(file = "filename.dat", center, convexity)
```

Boolean Operations

Group multiple shapes together into one shape:

```
union() { ... }
```

Subtract one or more shapes from an initial shape:

```
difference() { ... }
```

Draw the overlapping region of multiple shapes:

```
intersection() { ... }
```

Shape Transformations

Translate a shape according to a 2D or 3D vector:

```
translate([x, y, z])
```

Rotate a shape around each axis according to the angles defined by a vector:

```
rotate([x, y, z])
```

Rotate a shape a specific angle around the z-axis:

```
rotate(angle)
```

Scale a shape according to the scale factors defined by a 2D or 3D vector:

```
scale([x, y, z])
```

Resize a shape according to the dimensions defined by a 2D or 3D vector; optionally use auto to preserve the object aspect ratio in the unspecified dimensions:

```
resize([x, y, z], auto, convexity)
```

Reflect a shape according to the perpendicular vector of a symmetry plane passing through the origin:

```
mirror([x, y, z])
```

Multiply the geometry of all child elements with the given 4×4 affine transformation matrix:

```
multmatrix(matrix)
```

Change a shape's color according to a predefined color name or hexadecimal color value; optionally make the color (semi) transparent:

```
color("colorname | #hex", alpha)
```

Change a shape's color according to an RGB or RGBA vector. Each value in the vector ranges from 0 to 1 and represents the proportion of red, green, blue, and alpha present in the color.

```
color([r, g, b, a])
```

Move 2D outlines outward or inward by a given radius (for rounded corners) or delta + chamfer (for sharp or cut-off corners):

```
offset(r|delta, chamfer)
```

Create a 2D shape by projecting a 3D shape onto the xy-plane; when cut = true, create a 2D slice of the intersection of a 3D object and the xy-plane; optionally, when cut = true:

```
projection(cut)
```

Create a convex hull around one or more shapes:

```
hull() { ... }
```

Draw the Minkowski sum of multiple shapes:

```
minkowski() { ... }
```

Extrude a 2D shape into 3D with the given height along the z-axis; optionally center the shape at (0,0) or specify the convexity, twist, slices, and scale of the extrusion:

```
linear_extrude(height, center, convexity, twist, slices, scale)
```

Extrude a 2D shape around the z-axis to form a solid that has rotational symmetry:

```
rotate_extrude(angle, convexity)
```

Loops, Decisions, and List Comprehensions

Repeat a collection of shapes according to the start, step, and end (inclusive) values of a control variable:

```
for (var_name = [start:step:end]) { ... }
```

Draw the intersection of all the shapes generated by the for loop:

```
intersection_for(var_name - [start:step:end]) { ... }
```

Execute commands only if the Boolean test is true:

```
if (boolean_test) { ... }
```

Execute a collection of commands if the Boolean test is true; otherwise, execute alternate commands:

```
if (boolean_test) { ... } else { ... }
```

Generate a list of values according to a for loop:

```
list_var = [ for (i = range|list) func(i) ]
```

Generate a list of values according to a for loop, but only if the value causes a certain condition to be true:

```
list_var = [ for (i = ...) if (condition(i)) func(i) else ... ]
```

Generate a list of lists according to a for loop:

```
list_var = [ for (i = ...) let (assignments) func(...) ]
```

Other Shape Operations

Force the generation of a mesh even in preview mode:

```
render(convexity) { ... }
```

Inside a user-defined module, select the children specified by an index, vector, or range:

```
children(index | vector | range)
```

Modifier Characters

* Disables the drawing of a shape.
! Shows only a particular shape.
Highlights a shape in red for debugging purposes; highlighted shape will be rendered.
% Highlights a shape in gray; highlighted shape will not be rendered.

Special Variables

Mathematical Functions

sin(*ANGLE*)   Calculates the sine of an angle in degrees.

cos(*ANGLE*)   Calculates the cosine of an angle in degrees.

tan(*ANGLE*)   Calculates the tangent of an angle in degrees.

acos(*NUMBER*)   Calculates the arc (inverse) cosine, in degrees, of a number.

asin(*NUMBER*)   Calculates the arc (inverse) sine, in degrees, of a number.

atan(*NUMBER*)   Calculates the arc (inverse) tangent, in degrees, of a number.

atan2(*y, x*)   Two-value arc (inverse) tangent; returns the full angle (0–360) made between the x-axis and the vector [*x, y*].

abs(*NUMBER*)   Calculates the absolute value of a number.

sign(*NUMBER*)   Returns a unit value that extracts the sign of a value.

floor(*NUMBER*)   Calculates the largest integer not greater than the number.

ceil(*NUMBER*)   Calculates the next highest integer value.

round(*NUMBER*)   Calculates the rounded version of the number.

ln(*NUMBER*)   Calculates the natural logarithm of a number.

exp(*NUMBER*)   Calculates the mathematical constant e (2.718 . . .) raised to the power of the parameter.

log(*NUMBER*)   Calculates the base 10 logarithm of a number.

pow(*NUMBER, NUMBER*)   Calculates the result of a base raised to an exponent.

`sqrt(`*NUMBER*`)`   Calculates the square root of a number.

`rands(`*min, max, count, seed*`)`   Generates a vector of random numbers; optionally includes the seed for generating repeatable values.

`min(`*VECTOR | a, b, c*`)`   Calculates the minimum value in a vector or list of parameters.

`max(`*VECTOR | a, b, c*`)`   Calculates the maximum value in a vector or list of parameters.

`norm(`*VECTOR*`)`   Returns the Euclidean norm of a vector.

`cross(`*VECTOR, VECTOR*`)`   Calculates the cross-product of two vectors in 3D space.

Other Functions

`len(`*VECTOR|STRING*`)`   Calculates the length of a vector or string parameter.

`echo(`*STRING*`)`   Prints a value to the console window for debugging purposes.

`concat(`*VECTOR,VECTOR, ...*`)`   Returns a new vector that's the result of appending the elements of the supplied vectors.

`lookup(...)`   Looks up a value in a table and linearly interpolates whether there's no exact match.

`str(...)`   Converts all parameters to strings and concatenates.

`chr(`*NUMBER | VECTOR | STRING*`)`   Converts ASCII or Unicode values to a string.

`ord(`*CHARACTER*`)`   Converts a character into an ASCII or Unicode value.

`search(...)`   Finds all occurrences of a value or list of values in a vector, string, or more complex list-of-list construct.

`version()`   Returns the OpenSCAD version as a vector.

`version_num()`   Returns the OpenSCAD version as a number.

`parent_module(`*INDEX*`)`   Returns the name of the module `idx` levels above the current module in the instantiation stack.

`is_undef(`*VARIABLE*`)`, `is_list(`*VARIABLE*`)`, `is_num(`*VARIABLE*`)`, `is_bool(`*VARIABLE*`)`, `is_string(`*VARIABLE*`)`, `is_function(`*VARIABLE*`)`   Returns true if the argument is of the specified type.

`assert(`*expression*`)`   Will cause a compilation error if the expression is not true.

`let (`*variable = value*`) ...`   Assigns a value to a variable only in the following expression.

B

OPENSCAD VISUAL REFERENCE

This appendix is a quick visual reference for drawing, transforming, and combining the 3D and 2D shapes covered in this book. Paired with each screenshot is an example OpenSCAD statement that can be used to generate the image. In some cases, we've included a "shadow" object to illustrate what shapes looked like before an operation took place. Example code statements don't generate these shadow objects.

3D Primitives

Cuboid:

`cube([30, 20, 10]);`

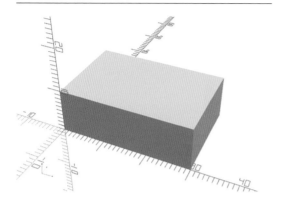

Centered cuboid:

`cube([30, 20, 10], center=true);`

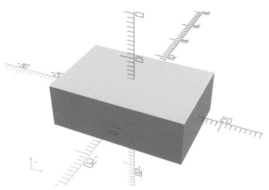

Sphere:

`sphere(10);`

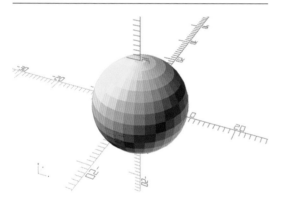

Smooth sphere:

`sphere(10, $fn=100);`

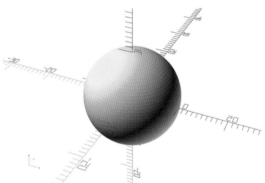

Cylinder:

`cylinder(h=20, r=5);`

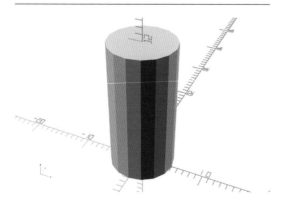

Cone:

`cylinder(h=20, r1=5, r2=0);`

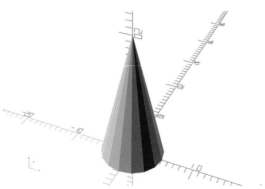

Centered smooth truncated cone:

```
cylinder(h=10, r1=3, r2=5, $fn=100, center=true);
```

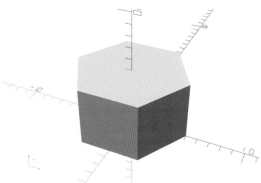

Regular prism:

```
cylinder(h=5, r=5, $fn=6);
```

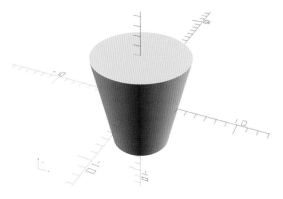

2D Shapes

Rectangle:

```
square([30, 20]);
```

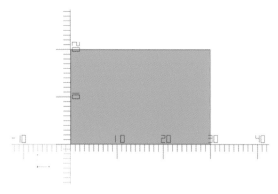

Centered rectangle:

```
square([30, 20], center=true);
```

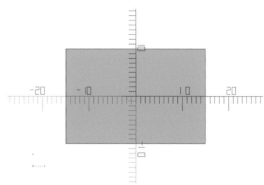

Circle:

```
circle(10);
```

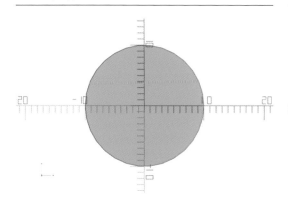

Regular polygon:

```
circle(10, $fn=5);
```

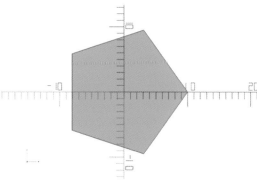

Irregular polygon:

```
polygon([[0,0], [10,0], [10,10], [5,10]]);
```

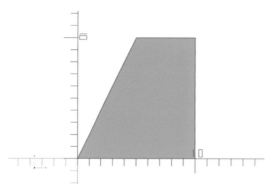

Text:

```
text("hello", font="Sans", size=20);
```

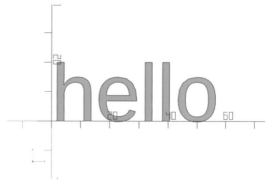

Combining Shapes

Subtracting from a shape:

```
difference() {
  sphere(10);
  translate([0,-15,0]) cube([15,30,15]);
}
```

Multiple subtractions from a shape:

```
difference() {
  sphere(10);

  cube([15, 15, 15]);
  cylinder(h=15, r=5);
}
```

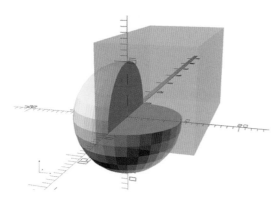

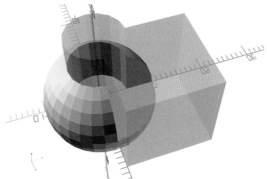

Intersection of two shapes:

```
intersection() {
  cube([10, 10, 10]);
  cylinder(h=15, r=5);
}
```

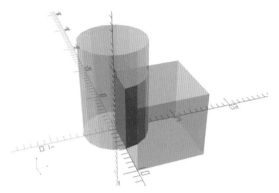

Subtracting from combined shapes:

```
difference() {
  union() {
      sphere(10);
      cylinder(h=30, r=5, center=true);
  }
  cube([10, 30, 10], center=true);
}
```

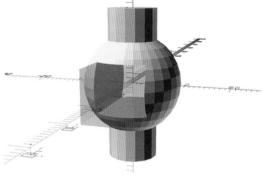

Convex hull:

```
hull() {
    sphere(10);
    cylinder(h=20, r=5);
}
```

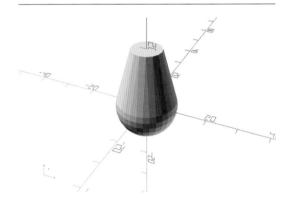

Minkowski sum:

```
minkowski() {
    sphere(10, $fn=50);
    cylinder(h=20, r=5);
}
```

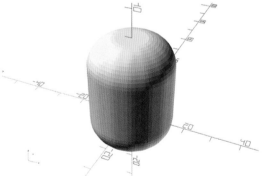

Transformations

Translation:

```
translate([5, 10, 0]) cube([5, 3, 1]);
```

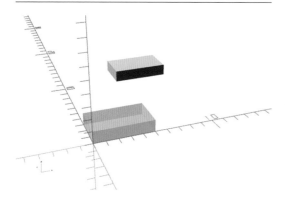

Rotation:

```
rotate([0, 0, 60]) cube([30, 20, 10]);
```

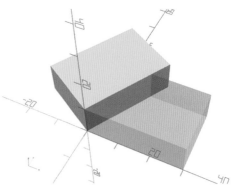

Reflection:

```
mirror([1, 0, 0]) translate([5, 0, 0])
  cylinder(h=1, r=5, $fn=5);
```

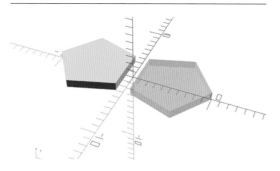

Resize dimensions:

```
resize([15, 20, 4]) sphere(r=5, $fn=32);
```

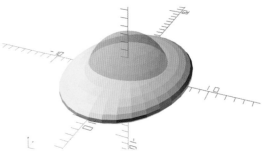

Extrude a 2D shape:

```
linear_extrude(height=10) {
    polygon([[0, 0], [10, 0],
             [10, 10], [5, 10]]);
}
```

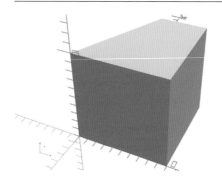

Rotate an extrusion of a 2D shape:

```
rotate_extrude(angle=180) translate([10, 0])
  circle(5);
```

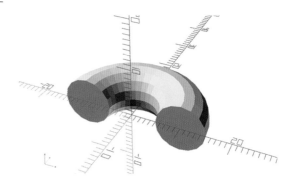

Loops

Repeat a shape:

```
for (x=[0:10:40]) {
    translate([x, 0, 0]) cube([5, 5, 10]);
}
```

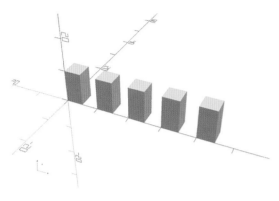

Vary characteristics of a repeated shape:

```
for (x=[0:1:4]) {
    h = x*5 + 5;
    translate([x*10, 0, 0]) cube([5, 5, h]);
}
```

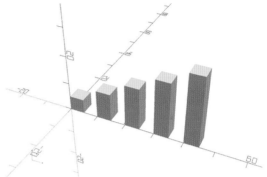

Repeat the repetition of a shape:

```
for (z=[0:15:45]) {
  for (x=[0:10:40]) {
    translate([x, 0, z]) cube([5, 5, 10]);
  }
}
```

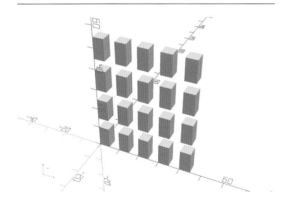

INDEX

Programming with OpenSCAD: A Beginner's Guide to Coding 3D-Printable Objects is set in New Baskerville, Futura, and Dogma. The book was printed and bound by Versa Printing in East Peoria, Illinois. The paper is 70# White Coated (Matte), which is certified by the Forest Stewardship Council (FSC).

The book uses a layflat binding, in which the pages are bound together with a cold-set, flexible glue and the first and last pages of the resulting book block are attached to the cover. The cover is not actually glued to the book's spine, and when open, the book lies flat and the spine doesn't crack.

Never before has the world relied so heavily on the Internet to stay connected and informed. That makes the Electronic Frontier Foundation's mission—to ensure that technology supports freedom, justice, and innovation for all people—more urgent than ever.

For over 30 years, EFF has fought for tech users through activism, in the courts, and by developing software to overcome obstacles to your privacy, security, and free expression. This dedication empowers all of us through darkness. With your help we can navigate toward a brighter digital future.

RESOURCES

Visit *https://nostarch.com/programmingopenscad/* for errata and more information.

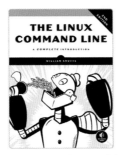
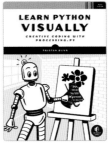

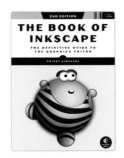